P9-DWA-304

LIBRARY IN A BOOK

GLOBAL WARMING

Douglas Long

Facts On File, Inc.

GLOBAL WARMING

Facts On File, Inc.
132 West 31st Street
New York NY 10001

Library of Congress Cataloging-in-Publication Data

Long, Douglas, 1967-
Global warming / Douglas Long.
p. cm.—(Library in a book)
Includes bibliographical references and index.
ISBN 0-8160-5137-2 (acid-free paper)
1. Global warming. I. Title. II. Series.
QC981.8.G56L66 2004
363 .738'74—dc21 2003003501

Facts On File books are available at special discounts when purchased in bulk quantities for businesses, associations, institutions, or sales promotions. Please call our Special Sales Department in New York at (212) 967-8800 or (800) 322-8755.

You can find Facts On File on the World Wide Web at http://www.factsonfile.com

Text design by Ron Monteleone
Graphs and diagrams by Sholto Ainslie

Printed in the United States of America

MP Hermitage 10 9 8 7 6 5 4 3 2 1

This book is printed on acid-free paper.

CONTENTS

PART III
APPENDICES

PART I

OVERVIEW OF THE TOPIC

CHAPTER 1

ISSUES IN GLOBAL WARMING

The issue of global warming first entered the general public's consciousness during the unusually hot summer of 1988, when James Hansen of the National Aeronautics and Space Administration's Goddard Institute for Space Studies warned that the Earth's temperature was steadily rising. Most scientists specializing in climate studies agreed that the trend was at least partially caused by carbon dioxide emissions from industrial activity.

The United Nations, alarmed by this and similar revelations, established the Intergovernmental Panel on Climate Change (IPCC) by the end of 1988 to assess the impacts of global warming and suggest strategies by which nations could curb carbon dioxide emissions. Subsequent research revealed that the 10 warmest years in recorded history (dating back 150 years) occurred during the 1980s and 1990s, and that the average global temperature rose by 0.6 degree Celsius (1 degree Fahrenheit) during the 20th century. The IPCC's First Assessment Report, issued in 1990, concluded that the amount of carbon dioxide in the atmosphere would double by the middle of the 21st century, resulting in the fastest rate of climate change in the last 10,000 years.

Dire consequences were predicted if something was not done to curtail carbon dioxide emissions on a massive scale. According to some scientists, melting glaciers and polar ice sheets would cause ocean levels to rise anywhere from 15 centimeters (6 inches) to 1 meter (3 feet) by the end of the 21st century, threatening the existence of coastal cities around the world. Higher temperatures would hasten the rate of water evaporation, increasing the frequency and magnitude of droughts. Global warming would threaten the existence of plants and animals unable to adapt or migrate quickly enough to keep pace with the rate of climate change. Finally, increased temperatures would maximize the breeding range of disease-bearing mosquitoes, spreading diseases such as malaria and dengue fever into areas where people lacked immunity.

Theories of global warming were not new to the late 20th century. In the 19th century, French scientist Jean-Baptiste-Joseph Fourier became the first person to envision Earth as a giant greenhouse whose atmosphere traps the radiant heat from the Sun, warming the planet and giving life to all organisms inhabiting its surface. About 70 years later, a Nobel Prize–winning chemist from Sweden, Svante August Arrhenius, became the first to suggest that the massive consumption of fossil fuels was capable of raising the temperature of the atmosphere.

Evidence for global warming was gathered throughout the 20th century. In 1938, Guy S. Callendar used data collected at 200 weather stations around the world to conclude that the Earth's temperature had risen 0.6 degree Celsius (1 degree Fahrenheit) during the previous 50 years. Measurements at Mauna Loa Observatory in Hawaii showed a global rise in atmospheric carbon dioxide of 11 percent between 1958 and 1997. Meanwhile, ice core samples from Antarctica and Greenland revealed an increasing trend in atmospheric carbon dioxide since the Industrial Revolution of the late 18th and early 19th centuries.

While there is little doubt that both global temperatures and atmospheric carbon dioxide have increased in the past 100 years, not all scientists agree on the extent of the correlation. Skeptics maintain that rising temperatures are the result of a natural climate variation, most likely a recovery from the deep cooling of the Little Ice Age. They also claim that global warming "hysteria" is based on poor interpretation of data, as well as overreliance on the faulty predictions of computer climate models. Furthermore, skeptics say that moderate warming would produce a more stable climate in which agriculture would thrive. These scientists counsel a "wait-and-see" political strategy, warning that more data is needed before costly carbon dioxide–cutting technologies are put into place.

Environmentalists are quick to claim that many global warming skeptics have accepted funding from oil and coal companies, that they represent a distinct minority opinion among the world's scientists, and that their arguments about misinterpreted data and faulty computer modeling, once valid, are now obsolete. Many economists are of the opinion that making a transition from fossil fuels to renewable energy will become costlier the longer it is delayed, and that the agricultural benefits of warmer temperatures will be limited to the Northern Hemisphere. Crops in the developing world, meanwhile, would be decimated.

The international community has taken steps to confront the problem of global warming. These include forming the IPCC, as well as drafting the 1992 United Nations Framework Convention on Climate Change (UNFCCC), which urged signatory nations to reduce greenhouse emissions to 1990 levels by 2000. Five years later, representatives from more than 150 countries, dissatisfied with the voluntary nature of the UNFCCC, gathered in Kyoto, Japan, to draft legally binding targets and timetables for reducing greenhouse gas emissions. A debate erupted between developing and wealthy nations over who should pay for the necessary technological transition, but an agreement was eventually reached that called for industrialized countries to cut emissions by an average of 5 percent below 1990 levels for the years 2008 through 2012.

Although the United States signed the Kyoto Protocol during the Clinton administration, President George W. Bush declared the agreement "fatally flawed" and refused to ratify it on the grounds that it would have a negative impact on the U.S. energy industry and the economy. Far from killing the international agreement, Bush's announcement prompted representatives from 178 nations to finalize many of the protocol's key points during a July 2001 meeting in Germany while U.S. negotiators stood by without contributing. The world community thus made it clear that it has resolved to confront potential global warming problems with or without the involvement of the United States, China, India, and other countries reluctant to contribute to the solution.

Even after the September 11, 2001, attacks on New York City and Washington, D.C., when President Bush was forced to temper his unilateral approach to international politics in order to find allies for his war on terrorism, the United States continued to stand virtually alone in refusing to confront climate change on a global level. Some social analysts questioned the wisdom of this approach, pointing out that environmental and political stability go hand in hand and that early casualties of a succession of ecological disasters could be individual liberties and governmental democracy, allowing both terrorism and despotism to thrive. Global warming skeptics, on the other hand, continued to claim that taking action would harm the U.S. economy, which would in turn destabilize the economic and political infrastructure of developing nations that relied on the United States for financial support. Whatever path the United States chooses to take on the issue of global warming during the 21st century, it is clear that policy decisions will be based not only on available scientific data but also on domestic and international politics. Whether this will prove beneficial or detrimental to the long-term environmental health of the planet remains to be seen.

THE GREENHOUSE EFFECT

In the early 1820s, Jean-Baptiste-Joseph Fourier—a French scientist who had already conducted groundbreaking research into heat diffusion that would later influence heat theory in the fields of physics, theoretical astronomy, and engineering—began to investigate the question of how the Earth stays warm enough to support the diverse range of living organisms on its surface. His work led him to speculate that the Earth's atmosphere acts as "a domed container made of glass, a gigantic bell jar formed out of clouds and invisible gases" that traps some of the solar energy reflected from the planet's surface, preventing it from escaping back into space. This trapped heat is reradiated down to the Earth's surface, keeping it warm enough to support life. Although he published his hypothesis in the article "General Remarks on the Temperature of the Terrestrial Globe and Planetary Spaces" (1824), the idea remained obscure until it was revived late in the 19th century.[1]

Scientists now recognize that the temperature of Earth's surface depends on a balance of incoming radiation from the Sun (solar radiation) and outgoing radiation from the surface of the Earth (terrestrial radiation). Only a small amount of the heat energy emitted from the planet's surface passes through the atmosphere back into space. The majority is absorbed by molecules of carbon dioxide, water vapor, methane, nitrous oxide, chlorofluorocarbons (CFCs), and ozone, collectively known as greenhouse gases because they are much more efficient at reflecting or trapping radiation than other gases. This trapped radiation contributes to the energy radiated back down to warm the Earth's surface and the lower atmosphere (see Appendix A: The Greenhouse Effect and Global Warming Trends).

Without greenhouse gases, the average temperature on Earth would be about 33 degrees Celsius (59 degrees Fahrenheit) colder and would be unable to support life as we know it.[2] Greater levels of greenhouse gases in the atmosphere, on the other hand, can lead to rising surface temperatures. In extreme cases, this is known as the runaway greenhouse effect. A well-known example of this effect occurs on the planet Venus, which receives twice the amount of solar radiation as Earth and possesses an atmosphere that consists of more than 96 percent carbon dioxide. The combined effect keeps the average surface temperature around 460 degrees Celsius (860 degrees Fahrenheit).

The high heat-absorbing property of greenhouse gases was first measured by English engineer John Tyndall, who published his findings in 1861. He speculated that a significant reduction in the level of atmospheric carbon dioxide could lead to another ice age, but he never considered the possibility that rising levels could result in higher global temperatures.[3]

This latter idea was left to Nobel Prize–winning Swedish chemist Svante August Arrhenius, who set out to answer the question, "Is the mean temperature of the ground in any way influenced by the presence of heat-absorbing gases in the atmosphere?" To do so, he created the first theoretical model to calculate the influence of carbon dioxide on Earth's temperature. Influenced by the work of Fourier, Tyndall, and U.S. astronomer Samuel P. Langley (who theorized that the absence of greenhouse gases would freeze the planet), Arrhenius published the paper "On the Influence of Carbonic Acid in the Air upon the Temperature of the Ground" (1896), in which he theorized that water vapor and carbon dioxide were responsible for warming the Earth's atmosphere. He coined the term "hothouse" to describe this effect. (It was later renamed "greenhouse effect" by other scientists.)

According to Arrhenius's mathematical calculations, a loss of 50 percent of the atmosphere's carbon dioxide would result in a temperature drop of 4 to 5 degrees Celsius (7.2 to 9 degrees Fahrenheit), enough to trigger another ice age. A doubling of carbon dioxide levels would lead to an average temperature increase of 5 to 6 degrees Celsius (9 to 11 Fahrenheit). Based on his belief that the consumption of fossil fuels inaugurated by the Industrial Revolution of the 18th and 19th centuries was adding massive amounts of carbon dioxide to the atmosphere, Arrhenius concluded that an increase in temperature was more likely than a decrease. Far from being alarmed, the chemist estimated it would take 3,000 years of burning fossil fuels for the amount of atmospheric carbon dioxide to double from its late-19th-century level. Well acquainted with Sweden's difficult winters, he also believed that such warming could only benefit humanity, writing that increased temperatures would "allow all our descendants, even if they only be those of the distant future, to live under a warmer sky and in a less harsh environment than we were granted." In 1908 he added, "By the influence of the increasing percentage of [carbon dioxide] in the atmosphere, we may hope to enjoy ages with more equable and better climates, especially as regards colder regions of the Earth, ages when the Earth will bring forth much more abundant crops than at present for the benefit of rapidly propagating mankind."[4]

FOSSIL FUELS AND THE INDUSTRIAL REVOLUTION

Scientists who study past climates, known as paleoclimatologists, have used evidence gathered from ice cores, tree rings, corals, and lake and ocean sediments to construct a general picture of the evolution of Earth's climate and weather patterns. According to these records, over the past 1 million years the Earth has experienced a pattern of eight repeated ice ages that lasted about 100,000 years and were interspersed with shorter, warmer interglacial periods of 8,000 to 40,000 years. These cycles, discovered by Serbian mathematician Milutin Milankovitch in the 1920s, are thought by most scientists to be driven by changes in the Earth's orbit around the Sun. The most recent interglacial period, known as the Holocene, began about 10,000 years ago and continues to this day. Although research has indicated temperature swings as large as 5 to 8 degrees Celsius (9 to 14.4 degrees Fahrenheit) over 1,500 years during this period, the Holocene appears to be by far the longest warm and stable period in the past 400,000 years.[5]

Records of annual average Northern Hemisphere surface air temperatures derived from tree rings and corals indicate a slight cooling trend from A.D. 1000 to 1900, interrupted by the Medieval Warm Period (A.D. 900 to 1300; also called the Little Climatic Optimum), an era of slightly warmer temperatures and increased rainfall that facilitated the expansion of agriculture in the Northern Hemisphere and allowed Viking explorers to reach Greenland and North America. This was followed by a resumption of declining temperatures in the form of the Little Ice Age (A.D. 1300 to 1900), a period during which average temperatures in the North Atlantic region dropped by about 1.5 to 2 degrees Celsius (2.7 to 3.6 degrees Fahrenheit), causing extensive glaciation and shorter agricultural growing seasons in the Northern Hemisphere.

Since 1900, however, direct measurements have indicated a rapid rise in surface temperatures. According to the IPCC, the 20th century was the warmest of the millennium, while the 1990s were the warmest decade and 1998 the warmest year of the millennium in the Northern Hemisphere. Although few doubt that temperatures recorded in the late 20th century were higher than those of previous centuries, some scientists argue that this occurred because the planet is coming out of the Little Ice Age.[6] Others believe that the data used to reconstruct the climate of the past 1,000 years are faulty, as they are almost exclusively based on information gathered from North American tree rings. These records, some scientists complain, are too localized, they ignore the 70 percent of Earth that is covered in water, and, since trees grow mostly in summer and during daytime, they fail to reliably measure full annual temperatures.[7] The consensus among climate specialists, however, is that rising temperatures since 1900 are closely linked with the massive burning of fossil fuels that began during the Industrial Revolution.

Fossil fuels are created from the remains of ancient plants and animals that proliferated 354 to 280 million years ago during the Carboniferous period, a division of the Paleozoic era. Since all organic matter is made of carbon, which is derived from carbon dioxide, the increasing abundance of life during that period removed large amounts of carbon dioxide from the atmosphere. When organisms die, they are normally attacked by bacteria and other microorganisms that cause rapid decomposition. However, when dead organisms fall into swamps, of which there was an abundance during the wet Carboniferous period, this process of decay is compromised by lack of oxygen. The wet, rotted, partially carbonized organic material, known as peat, is instead buried under successive layers of mud and sand. Over millions of years, heat and pressure compressed and consolidated it into layers of coal. Beginning with the Industrial Revolution, people began mining these organic remains and burning them for energy, thus rereleasing long-dormant stores of carbon dioxide back into the atmosphere.

Throughout the 18th century, coal increasingly replaced wood as the fuel of choice in the Western world. In 1712, Thomas Newcomen and John Calley built the first successful coal-powered steam engine, designed to pump water from inundated mines. Within a decade, it was being used all over Europe. Textile production also came to rely on coal, and Richard Arkwright's new cotton mills marked the 1760s as the beginning of the industrial factory system. In 1769, James Watt invented an improved steam pump that kicked the Industrial Revolution into overdrive. By the early 19th century, coal-powered steam engines were finding applications in train travel. By the middle of the century, more than 1,300 miles of railroad track crisscrossed the countryside of Britain. Each one of these major inventions, as well as countless minor ones, increased the amount of coal burned and the amount of carbon dioxide released into the atmosphere.[8]

The next energy breakthrough occurred in 1859 when Edwin Laurentine Drake struck oil in Titusville, Pennsylvania. The resulting well, which produced 10 to 35 barrels per day, nearly doubled the world's oil production. The area became a boomtown when other nearby wells began producing oil, and the world's first petroleum refinery was built the following year about one mile from the original well. Oil found a major application in the transportation sector in 1885 when Gottlieb Daimler built the first four-wheeled road vehicle, dubbed the "horseless carriage," powered by an internal combustion engine, which had first been developed by Belgian inventor Étienne Lenoir in 1860. A series of improvements led to the construction of motors that could be used to power boats, fire engine pumps, small locomotives, and zeppelin airships. In 1908, U.S. entrepreneur Henry Ford ensured that oil would become one of the most important fuel sources in the world when his Ford Motor Car Company, based in Detroit, Michigan, began mass production of the Model T, dramatically increasing use of the internal combustion engine.[9]

This massive increase of fossil fuel burning due to the Industrial Revolution elevated the amount of carbon dioxide in the atmosphere. In 1850, atmospheric levels of the greenhouse gas were measured at 280 parts per million (ppm). By the mid-1990s, levels had reached 360 ppm. During this same general period, from 1856 to 2000, global temperatures increased by approximately 0.6 degree Celsius (1 degree Fahrenheit). Before the 1980s, when the general population became aware of global warming, a handful of scientists worked to establish a definitive link between the two trends.

GROWING EVIDENCE OF GLOBAL WARMING

In 1934, British meteorologist Guy S. Callendar published the first article on the greenhouse effect in decades, "The Artificial Production of Carbon Dioxide and Its Influence on Temperature." Basing his findings on data collected at 200 weather stations around the world between 1880 and 1934, and having factored in the heat island effect (heat retention that occurs in large cities), Callendar revealed that the Earth's temperature had risen nearly 0.6 degree Celsius (1 degree Fahrenheit) since 1880. He linked this warming trend to carbon dioxide released by the combustion of fossil fuels and estimated an additional 1.1-degree-Celsius (2-degree-Fahrenheit) rise in average global temperatures over the following century. Like Arrhenius, Callendar believed such warming would benefit humans, writing, "In conclusion it may be said that the combustion of fossil fuel, whether it be peat from the surface or oil from 10,000 feet below, is likely to prove beneficial to mankind in several ways, besides the provision of heat and power." These added benefits included improved agriculture, a general increase in plant growth, and a delay in the return of glaciers.[10]

The steady rise in temperature detected by Callendar that began in the 1880s peaked around 1940, after which temperatures declined for nearly 40 years. By the late 1970s, most of the 0.6-degree-Celsius (1-degree-Fahrenheit) increase detected by the British meteorologist had disappeared. Some scientists theorized that the decline had been caused by Sun-blocking volcanic discharge, which was about double the average amount between 1945 and 1975.[11] Some skeptics, however, have pointed to the falling temperatures as evidence that global warming is a fallacy and that there is, at best, only a tenuous connection between global temperatures and atmospheric carbon dioxide levels.

Pioneering work in the measurement of carbon dioxide in the atmosphere was begun in the 1950s by U.S. chemist Charles Keeling. In order to research the question of whether carbon dioxide in the atmosphere was in balance with the carbon dioxide in the oceans, Keeling was forced to build a device capable of measuring atmospheric gas in parts per million (ppm). He based his design on a 1916 article describing the manometer, an instrument designed to measure small amounts of various gases. In 1955, Keeling began collecting air samples along the West Coast of the United States and, although scientists had previously assumed that carbon dioxide levels varied from place to place, continuously measured 315 ppm of carbon dioxide. The following year Keeling began working at the Scripps Institution of Oceanography in La Jolla, California, where he improved the accuracy of the manometer by a factor of 10.

In 1958, one of the new instruments was placed on Mauna Loa, a 4,104-meter (13,680-foot) volcano on the island of Hawaii. The data collected over the following decades have shown a steady increase in the levels of carbon dioxide in the atmosphere, an upward trend known as the Keeling curve. From the 1958 baseline of 315 ppm, carbon dioxide levels exceeded 320 ppm during the 1960s, 330 ppm during the 1970s, 340 ppm during the 1980s, and 360 ppm by the year 2000, a global rise of 11 percent and the highest levels of the past 420,000 years.[12]

Despite the anomaly created by the slight decrease in temperatures from 1940 to the late 1970s, global temperature, carbon dioxide concentrations, and carbon emission levels all exhibited noticeable upward trends throughout the 20th century when compared with the record of the past 1,000 years. By 1980, the temperature decline had come to an end, and the global average surface temperature increased by a total of 0.6 degree Celsius (1 degree Fahrenheit) during the 20th century. About half this rise has occurred since the general warming trend resumed in the late 1970s, and 17 of the 18 warmest years of the century occurred between 1980 and 2000. In 1998, the global temperature set a new record by a wide margin, exceeding that of the previous record year, 1997, by about 0.2 degree Celsius (0.3 degree Fahrenheit). Higher latitudes warmed more than equatorial regions, and nighttime temperatures rose more than daytime temperatures.[13]

Keeling was not the only scientist conducting research into carbon dioxide levels during the 1950s. Roger Revelle and Hans E. Suess both oceanographers with the Scripps Institution of Oceanography in La Jolla, California, investigated the role of the world's oceans in the carbon cycle. This cycle is the process by which carbon, in the form of carbon dioxide, is exchanged among natural reservoirs, or sinks, that are capable of absorbing and storing (sequestering) carbon released from another part of the cycle. Storage of carbon in such sinks—which include oceans, vegetation, and unburned fossil fuels—can partially offset the effects of anthropogenic carbon dioxide emissions. Revelle and Suess concluded that the oceans were not absorbing as much of the carbon dioxide emitted to the atmosphere as had been previously thought and that rising levels of carbon dioxide in the atmosphere could cause global temperatures to rise. They published their findings in 1957, writing that humans were "carrying out a large scale geophysical experiment of a kind that could not have happened in the past nor be reproduced in the future."[14]

Taken as a whole, the research conducted over the 140 years between the 1820s and the 1950s by Fourier, Tyndall, Langley, Arrhenius, Callendar, Keeling, Revelle, and Suess began to paint a picture of carbon dioxide emissions from the burning of fossil fuels exerting a powerful, if benign, influence on the climate of planet Earth. It would take experiments with computer models of the atmosphere to turn these isolated findings into the coherent yet controversial theory that came to be known as global warming.

COMPUTER MODELS OF THE ATMOSPHERE

On March 5, 1950, Jule Charney, John Von Neumann, and other scientists at the Meteorology Project in Princeton, New Jersey, launched the first computerized simulation of the weather. Von Neumann, a Hungarian-born mathematician who had moved to the United States in 1930, is credited with designing one of the first electronic computers, which he chose to apply to weather forecasting. Charney, a theoretical meteorologist who had been appointed director of the Meteorology Project in 1948, focused on the difficult problem of converting the physical properties that govern the behavior of the atmosphere into mathematical formulas that could be fed into a computer. The project revolutionized the use of computerized weather predictions and led to the development of general circulation models (GCMs, also called atmosphere-ocean general circulation models, or AOGCMs), which simulate the interaction of the Earth's atmosphere and oceans and the manner in which they change over time.

Despite Charney's and Von Neumann's success, their breakthrough also led them to believe that the behavior of the atmosphere, as well as the interaction between the atmosphere and the oceans, is much too complex for computers to accurately simulate and predict. This high level of ambiguity would later become one of the primary weapons in the arsenal of skeptics who argue that global warming is not the dire problem that many scientists believe.[15]

Application of computers to global warming studies began in 1958 when Joseph Smagorinsky, a scientist who had worked with Charney and Von Neumann at the Meteorology Project, recruited Japanese physicist Syukuro Manabe to work on the problem of accounting for the greenhouse effect in computer simulations of the atmosphere. One of Manabe's primary contributions to the project was to reformulate the general climate model to include the process of convection that occurs when greenhouse warming raises the temperature of the Earth's surface.

In 1967, Manabe and his colleagues at the U.S. federal government's Geophysical Fluid Dynamics Laboratory (GFDL) used this improved model to calculate that a doubling of preindustrial carbon dioxide levels would lead to a global average temperature increase of 2.2 degrees Celsius (4 degrees Fahrenheit). The following year, Manabe and co-researcher Richard Wetherald developed the first three-dimensional computer model of the entire Earth. The new simulation increased the predicted rise in average temperature that would accompany a doubling of carbon dioxide levels to 3 degrees Celsius (5.4 degrees Fahrenheit). The model also calculated an 8 percent increase in evaporation of water in the atmosphere. These findings were published by Manabe and Wetherald in 1975, the same year Charles Keeling released updated measurements of atmospheric carbon dioxide concentrations showing that levels were increasing. Taken together, the two articles alerted many scientists to the idea that human activities can affect global climate.[16]

These revelations were followed by a flurry of research projects that examined the question of rising temperatures, but global warming theorists also had to contend with the opposing hypothesis that the planet was heading toward another ice age. The 40-year cooling trend and expansion of glaciers, coupled with a series of unusually harsh winters during the latter half of the 1970s, convinced some scientists that a new ice age was imminent. Climatologists Reid Bryson and Stephen S. Schneider, among others, speculated that human-made pollutants were reducing the amount of sunlight reaching the Earth. This cooling theory was abandoned by most proponents in the late 1970s when global temperatures again began to rise. Schneider was among those who changed their mind and later supported the view that greenhouse emissions were causing the Earth to warm. Global warming skeptics have pointed to this reversal as evidence that the science of climatology is too uncertain to be used as the basis for policy decisions, but Schneider has maintained that the ability to change one's mind in the face of new evidence is the "ultimate test" of respectability for scientists.[17]

In 1975, even before the cooling trend had ended, Columbia University professor Wallace S. Broecker warned that a pronounced warming trend induced by carbon dioxide would become evident within a decade. The following year, researchers James Hansen and Yuk Ling Yung published a paper in the journal *Science* that identified methane and nitrous oxide as greenhouse gases. In 1979, a group of scientists that included David Keeling and Roger Revelle issued a report to the President's Council on Environmental Quality, concluding that "man is setting in motion a series of events that seem certain to cause significant warming of world climate unless mitigating steps are taken immediately." In October of the same year, a study by the National Academy of Sciences (NAS) panel on climate change, initiated by President Jimmy Carter's science adviser in 1976, warned that a doubling of atmospheric carbon dioxide could raise global temperatures by 1.5 to 4.5 degrees Celsius (2.7 to 8.1 degrees Fahrenheit). The panel warned that "a wait-and-see policy may mean waiting until it is too late" to avoid significant global warming.

GLOBAL WARMING POLICY

These findings and recommendations marked the beginning of global warming as a policy issue, a role that grew quietly throughout the 1980s.

In 1980, Council on Environmental Quality chairman Gus Speth urged policy makers to include concerns about carbon dioxide emissions and climate change in U.S. and global energy policies. In 1983, the U.S. Environmental Protection Agency (EPA) published "Can We Delay a Greenhouse Warming?" The report stated that global warming would result in conditions in which "agricultural conditions will be significantly altered, environmental and economic systems potentially disrupted, and political institutions stressed." Despite this warning, the report concluded that a ban on coal use, the only way to effectively slow the rate of global warming, would be politically and economically unfeasible.[18] In the same year, the U.S. National Academy of Sciences (NAS) published the report "Changing Climate," which found that the accumulation of greenhouse gases in the Earth's atmosphere could eventually raise global temperatures by 1.5 to 4.5 degrees Celsius (2.7 to 8.1 degrees Fahrenheit). The report went on to say that current evidence was not enough to warrant changes in energy policy.

Concerns were also beginning to grow on the international level. In 1980, representatives of the United Nations Environment Programme (UNEP), the World Meteorological Organization (WMO), and the International Council of Scientific Unions (ICSU) met in Villach, Austria, to analyze the scientific basis of claims that carbon dioxide emissions were contributing to global warming. Attendees concluded that climate change was a major environmental issue but recommended additional research before a management plan to control emissions was developed. In October 1985, the same bodies convened again in Villach bringing together scientists from 29 nations. The conference report warned that "some warming . . . now appears inevitable" regardless of future actions and recommended consideration of a global treaty to address climatic change. A report issued by UNEP in April 1988 titled "Development Policies for Responding to Climate Change" claimed that global warming would outpace the environment's ability to adapt to rising temperatures. The following month, delegates from 46 countries convened in Toronto, Ontario, at the International Conference on the Changing Atmosphere and suggested a reduction of global carbon dioxide emissions by 20 percent from 1988 levels by the year 2005. Norwegian prime minister Gro Harlem Brundtland called for a global convention on the greenhouse effect. Many believed such a treaty could be modeled on the Montreal Protocol, drafted and signed in 1987 in response to the discovery that human-made chlorofluorocarbons (CFCs), rare but potent greenhouse gases used in refrigeration, were destroying the atmospheric ozone layer. Signatories to the Montreal Protocol were obliged to adopt legislation and policies aimed at reducing activities likely to have an adverse effect on the ozone layer, including the gradual phase-out of CFCs and other ozone depleting substances (ODS).[19]

If the growing concern about global warming occurred below the radar of the average U.S. citizen, James Hansen of NASA's Goddard Institute of Space Studies made sure everyone in the nation knew about the problem by the end of the 1980s. In 1981, he published a study showing that the Earth's temperature had risen until 1940, cooled from 1940 to 1975, and began rising again during the mid-1970s. The net temperature change had been a warming of 0.4 degree Celsius (0.7 degree Fahrenheit). He went on to predict warming in the 21st century of "almost unprecedented magnitude" that could melt the West Antarctic ice sheet and cause the global sea level to rise by 4.5 to 6 meters (15 to 20 feet). At Senate hearings held in June 1986, Hansen predicted a warming of 1.7 degrees Celsius (3 degrees Fahrenheit) over 30 years if current levels of carbon dioxide emissions continued. As a result, Senator John Chafee asked the EPA and the Office of Technology Assessment to develop policy options for stabilizing greenhouse gas concentrations in the atmosphere. Congress responded by setting aside extra money in the EPA's budget to conduct climate change research.

The lid was blown off the great secret of global warming on June 23, 1988, when Hansen again testified before the U.S. Senate Committee on Energy and Natural Resources. On that oppressively hot and humid day in Washington, D.C., the well-respected scientist made the bold, unprecedented statement that "global warming is now large enough that we can ascribe with a high degree of confidence a cause and effect relationship to the greenhouse effect." He added dramatically, "It is already happening now."[20] The extent to which Hansen's testimony propelled the issue of global warming into the public arena and into the forefront of international politics became evident two months later when U.S. presidential candidate George H. W. Bush saw fit to address global warming in his campaign. He seemed to promise action on the issue when he stated that "those who think we are powerless to do anything about the greenhouse effect forget about the 'White House effect'; as president, I intend to do something about it."[21]

On the international level, the United Nations was sufficiently alarmed by warnings about global warming to form the Intergovernmental Panel on Climate Change (IPCC) under the auspices of WMO and UNEP in November 1988. Based in Geneva, Switzerland, this group of 2,500 prominent climate scientists was given the task of assessing the science on climate change, including causes, impacts, and possible responses. For some, the IPCC became the undisputed global authority on all matters pertaining to climate change. For others, the organization came to represent everything that was wrong about the theory of global warming.

INTERGOVERNMENTAL PANEL ON CLIMATE CHANGE

The IPCC seeks to produce a concrete consensus on the issue of climate change by giving scientists control of its operations while allowing its work to be reviewed and commented on by other scientists, government officials, and special interest groups, from environmentalists to representatives of the fossil fuel industry.[22] Reports, each of which includes a detailed summary of its findings for policy makers, are prepared by three independent working groups: Working Group I focuses on the scientific basis of climate change; Working Group II analyzes impacts and adaptation strategies, and explores the vulnerability of socioeconomic and natural systems to climate change; and Working Group III addresses the issue of mitigating greenhouse gas emissions.

The IPCC released its First Assessment Report in 1990. Researchers scrutinized all available temperature studies and reiterated the familiar observation that the average surface temperature of the planet had increased by about 0.6 degree Celsius (1 degree Fahrenheit) over the preceding century. Although the scientists involved were unable to determine whether this increase was caused by human-induced climate change or natural variability, they did conclude that a doubling of pre–Industrial Revolution carbon dioxide levels would likely occur near the middle of the 21st century, causing an increase in average global temperatures of as much as 4.5 degrees Celsius (8.1 degrees Fahrenheit).[23] The 1990 report provided the basis for formal negotiations of the United Nations Framework Convention on Climate Change (UNFCCC) at the Earth Summit held in Brazil in 1992.

The IPCC's Second Assessment Report was published in 1995, providing a new benchmark of international scientific consensus that became the basis for negotiating the 1997 Kyoto Protocol. In the five years between the IPCC's two reports, scientists greatly increased their ability to distinguish between natural and human influences on the climate, in particular by accounting for the release of sulfate aerosols (which have a cooling effect) in computer simulations of the climate. These calculations resulted in a better match between simulations of climate change and actual changes, an improvement that led the report's authors to conclude that "the observed warming trend is unlikely to be entirely natural in origin" and that "the balance of evidence suggests that there is a discernible human influence on global climate."[24]

Several of the 1995 report's findings solidified the divide between the IPCC and global warming skeptics. The report reaffirmed the 0.6-degree-Celsius temperature rise over the preceding century but refined the statistic by adding that more than half the warming since 1900 had occurred in the most recent 40 years. This contradicted assertions by many skeptics that most of the warming took place before 1940. The IPCC also concluded that readings taken by satellites and weather balloons showed a slight overall cooling of the troposphere (the region of the lower atmosphere that extends from the Earth's surface up to a height of about 10 kilometers, or 6 miles) after 1979. Skeptics cited this as proof that no recent warming had occurred, but the IPCC claimed that adjusting these records for the short-term effects of volcanic eruptions and warming caused by El Niño resulted in a slight warming, although not as large as that shown by surface data. El Niño (Spanish for "Christ child," so called because it usually occurs around Christmas) is a climatic phenomenon characterized by warmer-than-normal sea-surface temperatures in the tropical eastern Pacific Ocean off the coast of Peru. Occurring at irregular intervals every two to seven years and generally lasting 12 to 18 months, El Niño can influence rainfall patterns that lead to floods and droughts around the world, sometimes to devastating effect. Weather conditions related to the strong 1982–83 event (including droughts, floods, forest fires, and storms) killed 2,000 people around the globe and caused $13 billion of damage. El Niño has also adversely affected the South American fishing industry, as warmer top waters can prevent nutrients needed by fish from rising to the surface from colder depths. Although climate models have indicated that rising global temperatures will likely bring about an increase in the types of extreme weather associated with a strong El Niño, researchers have been unable to forge a direct link between global warming and the El Niño events of 1982–83 and 1997–98, which were more intense than those previously recorded.

The IPCC noted that most of the warming during the 20th century had occurred in the Northern Hemisphere, that nights had warmed more than days, and that winter had warmed more than summer. The fact that very cold temperatures had become less frequent while hot nights in the summer had become more frequent matched predictions made by the latest computer models. Skeptics, however, used this news of warmer nights and winters to argue that climate change would be largely beneficial to humankind.[25]

Improvements in the understanding of climate change led the authors of the Third Assessment Report, released in 2001, to conclude that "an increasing body of observations gives a collective picture of a warming world and other changes in the climate system." These included widespread decreases in snow cover and ice extent, as well as a sea level rise of 10 to 20 centimeters (4 to 8 inches), during the 20th century. The report reaffirmed the conclusion that the 1990s were likely the warmest decade, and 1998 the warmest year, since direct recording of temperatures began in the 1860s. Based on measurements in the Northern Hemisphere, scientists found that the average global surface temperature rose more during the 20th century than during any other century of the past 1,000 years.[26]

The IPCC went on to assert that between 1750 and 2001, atmospheric carbon dioxide concentrations increased 31 percent, more than half of this increase having occurred in the second half of the 20th century. At the time of the report, concentrations were the highest in the past 420,000 years, and probably the last 20 million years. About 75 percent of the anthropogenic (caused by humans) carbon emissions of the 20 years leading up to the Third Assessment Report were from fossil fuel emissions, while the remainder came from the destruction of carbon sinks in the form of deforestation and other types of land-use change. From all this, the IPCC concluded that while natural processes had made small contributions to warming of the previous 100 years, "there is new and stronger evidence that most of the observed warming of the last 50 years is attributable to human activities."[27]

The Third Assessment Report not only studied the past to determine patterns of human-induced climate change but also looked ahead in an attempt to predict how these patterns would play out and affect humans in the future. The IPCC expects emissions of carbon from the combustion of fossil fuels to be the primary influence on future carbon dioxide levels, projected to increase from 360 parts per million (ppm) measured in the 1990s to 540 to 970 ppm by 2100. At current rates, the study projects, temperature will increase by 1.4 to 5.8 degrees Celsius (2.5 to 10.4 degrees Fahrenheit) by 2100, a rate much larger than that experienced in the 20th century and likely to be without precedent in the last 10,000 years. Furthermore, average sea level is projected to rise by 9 to 88 centimeters (3.5 to 34.3 inches). Snow cover, sea ice, and glacier extent are all expected to decrease, while intense precipitation events in the Northern Hemisphere are "very likely" to increase. Even if greenhouse gases are stabilized, the study warns, climate change is likely to persist for many centuries, with surface temperatures and sea levels continuing to rise in response to past emissions.[28]

PROJECTED IMPACTS OF GLOBAL WARMING

While the IPCC strives to clarify the scientific basis of global warming, public discussions on the topic tend to focus not on research but on the potential impacts of higher global temperatures. These concerns can be divided into five major areas: oceans, freshwater resources, agriculture and food supply, forests, and human health.

Oceans

The most prominent ocean-related threat from global warming is sea level rise caused by the inflow of water from melting glaciers and polar ice sheets, and by thermal expansion (warmer water takes up more space than cooler water). Nearly 50 percent of the world's population lives in coastal zones. In the United States, the figure is 53 percent, with more people moving into the 17 percent of land on the coastal zone each year.[29]

Fears of sea level rise have been heightened by such spectacular events as the rapid disintegration in early 2002 of a Rhode Island–size portion of the Larsen B Ice Shelf on the Antarctic Peninsula, which has warmed about 2.2 degrees Celsius (4 degrees Fahrenheit) since the 1950s, a rate much faster than the global average. The massive shelf, thought to have existed since the end of the last major ice age about 12,000 years ago, lost 3,250 square kilometers (803,088 acres) of ice over a 35-day period starting in January 2002. Although some scientists have no qualms about attributing the partial disintegration of the Larsen B Ice Shelf to global warming, others point out that while the Antarctic Peninsula has warmed in recent decades, other areas of the Antarctic experienced significant cooling trends during the same period. Most researchers do agree on one thing: The climate of the region is complex and poorly understood.[30]

Areas vulnerable to rising sea levels include coastal cities such as New York and New Orleans; delta regions in Bangladesh, Egypt, the United States, and elsewhere; the Netherlands; and low-lying islands in the Pacific and other oceans. In Bangladesh, where about 7 percent of the habitable land is less than 1 meter (3 feet) above sea level, substantial amounts of good agricultural land could be lost. This would constitute a serious blow to a nation in which half the economy is based on agriculture and 85 percent of the population relies on agriculture for their livelihood.

Higher seas are likely to increase damage from storm surges and accelerate coastal erosion, which could threaten coastal development, transportation infrastructure, and tourism. Intrusion of saltwater into fresh groundwater sources could devastate supplies of water used for drinking and agriculture. Another potential impact is loss of estuaries, wetlands, and mangrove swamps. These areas contain an abundance of biological diversity, supplying humans with more than two-thirds of the fish caught for consumption. Many birds and animals rely on wetlands and coastal swamps for part of their life cycles. Rising sea levels could cause wetlands in some areas to extend inland, possibly at the expense of agricultural land, but where flood walls and other human constructions have been built, wetlands will simply disappear beneath rising waters.[31] The most alarming prospect of sea level rise is the possibility that low-lying island nations, particularly those situated on coral atolls, could disappear altogether.

Developed nations like the United States and the Netherlands may possess the economic and technological resources to protect themselves from sea level rise by building walls and other constructs to divert water, but developing countries, including Bangladesh and small island nations such as the Maldives in the Indian Ocean and the Marshall Islands in the Pacific, will likely be at the mercy of the advancing waters.

Coral reefs have also been adversely affected by warming seas. Nearly 500 million people around the world live within 100 kilometers (62 miles) of coral reefs, which provide a rich economic resource for fisheries, recreation, tourism, and coastal protection. Goods and services from reefs were valued in 1997 at $375 billion per year.[32] In addition, coral reefs are one of the largest global storehouses of marine biodiversity. In the United States, coral reefs play a major role in the environment and economies of Florida and Hawaii, as well as most U.S. territories in the Caribbean.

Corals have difficulty adapting to warming water, putting them under stress and causing them to expel the algae that live within them. This leaves corals white, or "bleached," and renders them susceptible to colonization by new, often deadly, algae and infections. In 1998, the warmest year on record, bleaching damaged huge expanses of coral around the world and sharply increased the percentage of damaged reefs. In some regions, as much as 70 percent of the coral may have died in a single season. According to the Global Coral Reef Monitoring Network (GCRMN), by 2000 about 27 percent of the world's coral reefs were effectively lost, compared with 10 percent in 1992. GCRMN estimated that without "urgent management action" to stem the damage, 40 percent of the reefs will be lost by 2010.[33]

Climate change is also very likely to put stress on fisheries by altering the distribution and abundance of major fish stocks around the world. In the United States alone, recreational and commercial fishing are estimated to contribute $40 billion per year to the economy. Long-term effects of global warming may include poleward shifts in the distribution of marine populations. The elevated sea surface temperatures that accompanied the 1997–98 El Niño are thought to be responsible for a number of unusual events, including a decrease in the abundance of squid and a widespread increase in the number of sea lion pup deaths off the coast of California, as well as poor salmon returns in Alaska. Some researchers worry that warming water will impact the supply of plankton, affecting, in turn, the life cycles of whales and other marine animals that rely on plankton for food.[34]

The 1997–98 coral losses and fishery changes have frequently been connected with El Niño, a climate phenomenon associated with temporary increases in water temperature and extreme weather events. Many scientists have questioned whether recent El Niños are entirely natural in origin or have been rendered more potent by global warming. El Niños normally occur every three to seven years and last 12 to 18 months. Computer climate models predict that global warming should bring about an increase in such extreme weather phenomena, and, indeed, since 1976 El Niños have occurred with a frequency unmatched in the previous century. The longest El Niño on record lasted from 1991 through mid 1995.[35]

Freshwater Resources

Many scientists believe that global warming is likely, in many regions, to alter the distribution of freshwater, an important resource that supports human activities and ecosystems. Increased temperatures will mean that evaporation will occur at a higher rate, promoting cycles of intense drought and flood. This occurs because a warmer atmosphere with higher evaporation rates makes it difficult for water molecules in air to stick together to form raindrops, thereby lengthening drought periods. During this time, however, more water vapor than usual will gather in the air. When a weather system finally triggers a rainstorm, the extra water vapor may lead to more intense rainfall and flooding. These floods have the potential to cause massive dislocations and damage in some regions. Droughts, meanwhile, may impact agriculture, water-based transportation, and ecosystems. A 1995–96 drought in the agricultural regions of the southern Great Plains resulted in about $5 billion in damages.[36] In 2002, a dry spell that caused water shortages and crop losses settled across 14 states along the East Coast and 14 more in the Rocky Mountains, the Great Plains, and the desert Southwest, about twice the area affected by severe drought in an average year. Some researchers believe that warming of the Indian Ocean over the past 50 years, possibly caused by greenhouse gas emissions, has intensified a region of low pressure in the North Atlantic, preventing storms from moving south. This has led to warmer winters and decreased precipitation in Europe and the eastern United States.[37] Droughts in arid and semi-arid regions, where long-term storage of water is often not an option, can have devastating and deadly effects on the populace.

A warming climate is likely to decrease snowpack, a natural water storage system in mountainous regions around the globe. Although precipitation is likely to increase in some regions, more of it will fall in the form of rain, causing the snow pack to develop later and melt earlier. As a result, peak stream flows will occur earlier in the spring, reducing summer flows in streams and rivers. This will increase competition for water supplies, not only between communities, but also between such needs as agriculture and conservation.

Water quality will also suffer. More frequent heavy precipitation events will likely flush more sediments and agricultural contaminants into rivers and lakes, degrading water quality. Excess nutrients from agricultural fertilizers, washed into lakes by heavy rains, combined with warmer water can cause algae blooms on the lake surface. These blooms deplete the ecosystem of oxygen, thereby harming other organisms. Flooding can also overload wastewater systems, sewage treatment facilities, and landfills, increasing the risk of contamination of drinking water supplies. Reduced summertime stream flows, on the other hand, can reduce water quality by increasing salinity.[38]

Agriculture and Food Supply

The vulnerability of water supplies to climate change can be extended to vulnerability in the growing of crops and production of food. Changes in precipitation type, timing, frequency, and intensity will affect crop yields, and warmer temperatures may cause explosions in insect populations, leading to increased use of pesticides to control them.

Global warming is expected to affect countries differently. Production in developed countries with relatively stable populations, such as the United States, may increase, whereas crop yields are likely to decrease in developing countries with increasing populations. Many of these countries are located in arid and semi-arid regions, where drought and famine already pose challenges to agriculture. The problem is expected to get worse, and the disparity between developed and developing nations will likely increase as global temperatures rise.

Although agriculture is highly dependent on climate, it is also enormously adaptive. Current knowledge of conditions required by different plant species, coupled with expertise in genetic manipulation, should allow crops to be matched to new climatic conditions. The negative effects of global warming may also be compensated for to some extent by longer growing seasons in some regions and increased productivity due to the fertilization effect of carbon dioxide. Higher concentrations of carbon dioxide stimulate photosynthesis and reduce water losses from plants, causing crop yields to rise. Crops that respond favorably to this effect (such as wheat, rice, barley, oats, and potatoes) generally experience yield increases of 15 to 20 percent with a doubling of atmospheric carbon dioxide. Corn, sorghum, sugarcane, and many tropical grasses are less responsive, with a doubling of carbon dioxide concentration yielding increases of 5 percent. Although skeptics often use these statistics as an example of the benign effects of global warming, the fertilization effect will be of negligible importance in areas where sufficient water supply and soil nutrients are not available.[39]

Forests

Organisms are accustomed to certain levels of rainfall and temperature provided by the ecosystem in which they live. If the environment changes, organisms are forced to adapt or migrate. Climate changes spurred by global warming, however, are expected to occur over a matter of decades, much too quickly for many species to respond or migrate. Scientists predict that climate zones may shift northward 160 to 560 kilometers (100 to 350 miles) and upward in altitude 150 to 540 meters (500 to 1,800 feet), which may eliminate habitats of alpine and subalpine trees that have nowhere to migrate. (Tree migration involves the expansion or shrinkage of entire forest zones over large areas and thousands of years in response to changes in climate. Plant migration caused by rising temperatures occurs as new generations of trees grow from seeds spread to the warming North by birds, wind, and other means.) Other plant and animal species may be rendered extinct if potential migration routes have been blocked by human development.[40]

Vulnerability to global warming varies greatly from species to species. It is obvious, for example, that birds can migrate more quickly than trees, which must rely on reproduction and long growth periods to shift their range. Forests are therefore likely to be the component of natural ecosystems most affected by climate change. As with agriculture, any benefits derived from the fertilization effect of elevated carbon dioxide may be undermined by inability to adapt to new temperature and precipitation patterns.

Forests are an important global resource, providing wildlife habitat, clean air and water, cultural and aesthetic value, recreational opportunities, and products that can be harvested, such as timber and fuelwood, wild game, mushrooms, and berries. They also act as a large carbon sink: 80 percent of aboveground and 40 percent of belowground terrestrial carbon is stored in forests. Widespread destruction of trees could significantly increase the amount of carbon dioxide in the atmosphere.[41] Forestry experts fear that global-warming-induced drought may increase the danger of wildfires, while others believe that rising temperatures may increase insect reproduction, further devastating vulnerable forests. South-central Alaska's spruce bark beetle plague of the 1990s, which affected 4 million acres, the biggest single insect kill of trees ever recorded in North America, was attributed by some scientists to the effects of global warming. An unbroken string of abnormally warm summers beginning in 1987 allowed beetles to grow to maturity in one year instead of two, doubling the hit on the forest the next spring, and allowed the plague to continue indefinitely instead of the normal one or two years.[42]

Declines in the health of forests will also lead to declines in the health of species that rely on them. Some species have already begun to adjust their ranges to ensure survival. Camille Parmesan, a graduate student at the University of Texas at Austin, began a rigorous, four-year field study of the range of the Edith's checkerspot butterfly *(Euphydryas editha)* in 1992. Her research showed that the butterfly's range had moved north and into higher altitudes, in accordance with what would be expected according to current global warming data. Animal species that are less mobile may not be able to adapt as quickly, if they can adapt at all.

Human Health

Human health is intricately bound to weather and environment. Many of the factors that lead to a deteriorated environment—pollution of the atmosphere, polluted or inadequate water supplies, and poor soil (resulting in poor crops and inadequate nutrition)—can also lead to poor health, and the effects of these factors may be exacerbated by climate change.[43]

With the projected increase in the severity and frequency of heat waves will likely come an increase in ailments associated with heat stress. Higher temperatures, combined with increasing amounts of fossil-fuel-based pollutants, will also likely lead to rising concentrations of ground-level ozone, which is known to aggravate respiratory disorders such as asthma. Death rates, particularly among the very young, the very old, and the poor, can double or triple during days of unusually high temperatures and humidity. A five-day heat wave in Chicago, Illinois, in 1995 brought maximum temperatures that ranged from 34 to 40 degrees Celsius (93 to 104 degrees Fahrenheit). The number of deaths during this period increased by 85 percent over the number during the same period of the preceding year. Officials recorded at least 700 deaths above what would normally be expected during moderate weather in the same period.

The positive effects of warmer temperatures may include a reduction in cold-weather deaths, although researchers lack sufficient understanding of the relationship between winter weather and mortality to determine whether milder winters will offset the increase in heat-related deaths during the summer. People can often adapt to hot conditions by using air conditioning and increasing their intake of fluids. These options, however, are less likely to be available to poor people or citizens of arid regions where water is scarce.

A warmer world may also see an increased spread of disease. Many insect carriers of disease thrive in warmer and wetter conditions, promoting the spread of malaria, yellow fever, dengue fever, and viral encephalitis. As noted earlier, extreme weather conditions such as floods may contribute to the spread of waterborne disease by overloading wastewater systems, sewage treatment facilities, and landfills. Increased runoff can also transport contaminants from agriculture and mine tailings (residue) into supplies of drinking water.[44]

GLOBAL WARMING SKEPTICS

The concerns enumerated in the previous section are taken seriously by respected climate specialists and government officials around the world. In-depth discussions in which a future characterized by such problems seems assured can be found everywhere from IPCC documents to policy statement of the U.S. government. However, not everyone agrees that global warming poses a threat to the planet and to humanity. A small but vocal group of skeptics has challenged the methods and conclusions of the IPCC since the release of the First Assessment Report in 1990. Some skeptics, warning that mitigation of greenhouse gas emissions would be too costly and would cause great harm to the global economy, believe that more emphasis should be placed on adaptation rather than prevention. Others maintain, like Svante August Arrhenius and other early theorists, that the effects of global warming will be benign, providing warmer winters, longer agricultural growing seasons, and increased plant growth due to the fertilizing effects of carbon dioxide. Still other skeptics say that global warming is not occurring at all, that the observed rise in temperature during the 20th century was the result of natural variation or even the result of faulty data. Among the more prominent skeptics are Robert C. Balling, Jr., director of the Office of Climatology and an associate professor of geography at Arizona State University; Richard Lindzen, Sloan Professor of Meteorology at the Massachusetts Institute of Technology (MIT) and a member of the National Academy of Sciences; Patrick J. Michaels, a research professor of environmental sciences at the University of Virginia and a fellow of the Cato Institute; Frederick Seitz, president of the National Academy of Sciences during the 1960s, chairman of the George C. Marshall Institute, and president emeritus of Rockefeller University; and S. Fred Singer, professor of environmental sciences at the University of Virginia.

Singer has articulated the general stance of many global warming skeptics through the publications of the Science and Environmental Policy Project (SEPP), a think tank he founded in 1990 to prove that the Earth has not experienced and will not undergo significant warming, and that any small rise in global temperature is more likely to benefit humans than harm them. Singer contends that although there is little doubt that global temperatures increased in the 20th century, most of this occurred well before greenhouse gas emissions increased appreciably. He also points out that the global temperature record of the 20th century shows both warming and cooling. This has led many skeptics to believe that climate change, rather than being definitively connected to anthropogenic causes, is likely the result of natural variation, such as a recovery from the deep cooling of the Little Ice Age, natural fluctuations caused by complex interaction between atmosphere and oceans, and variations of solar radiation. Climate change, Singer maintains, is not a new phenomenon, as evidenced by ice core studies that have revealed the existence of 17 ice age cycles in the past 2 million years.

One of Singer's main gripes about the IPCC, which he says does not submit its report chapters to peer review in the generally accepted sense despite

IPCC claims, is its reliance on faulty and poorly interpreted data. One of the primary points of contention is the accuracy of general circulation models (GCMs), which Singer and other skeptics say are not sufficiently sophisticated to match current weather observations, let alone predict future climate. The interaction between the atmosphere and the oceans is much too complex to feed into a computer and expect even a rough approximation of how they work. A faithful model of all the important factors in the climate system, according to skeptics, would involve representing everything from the entire planet down to individual dust particles, a task that current computers cannot handle. Many important climate processes, including clouds and ocean convection, occur on a scale much too small to be adequately modeled by the 250 kilometer by 250 kilometer by 1 kilometer grid into which GCMs typically divide the Earth's atmosphere. Skeptics are quick to point out that despite the inherent limitations of such climatic computer model simulations, all of the IPCC's global warming predictions are based on them.[45]

GCMs, according to Singer and others, tend to exaggerate the potential effects and dangers of global warming. Historically, warm periods have been beneficial for human populations, while cold periods have brought widespread disaster in the form of crop failure and disease. Instead of a doomed planet, skeptics tend to foresee one in which agriculture will thrive due to fewer frosts, extended growing seasons, and the fertilizing effects of carbon dioxide. Warming is expected to lead to a reduction in severe storms and may even lower rather than raise sea levels, because more evaporation from oceans would increase precipitation and thereby thicken the ice caps of Greenland and Antarctica. Spread of insect-borne tropical disease will not be an issue since the primary determinants of such diseases are insect control and public health strategies, both of which can easily be adapted to deal with the effects of increased insect populations.[46]

Skeptics also maintain that the effects of human-induced climate change in the 21st century will be relatively minor when compared to other prominent issues, such as population growth, economic growth, and adaptation to technological changes. Convinced that climate change will be a minor problem, skeptics counsel adaptation, coupled with a "no regrets" policy of conservation and efficiency improvements, as their preferred solution to global warming.[47] International global warming treaties such as the UNFCCC and the Kyoto Protocol are doomed to failure partly because researchers have made no attempts to define what constitutes a "dangerous" level of atmospheric carbon dioxide; therefore the goals of the treaties are arbitrary and "scientifically undefined."[48] More importantly, attempts to control greenhouse gas emission rates at this point are unrealistic and will have "disastrous economic consequences."[49] Energy or carbon taxes would "distort the economy, lower economic growth, raise consumer prices, lower standards of living, and destroy jobs" not only in industrialized nations but also in the majority of developing nations that use little energy.[50]

Singer does not discount the possibility of future emission reductions, if a time comes when such curbs can be accomplished without damaging the economy. Among his reasons to postpone such mitigation strategies are that existing capital stock should be allowed to wear out naturally before it is replaced; that because of the time value of money, deferred investments in capital equipment are preferable; that new technologies will be developed, whether higher efficiency or renewable energy, that will permit a lower-cost solution; and that scientists may learn more about climate science and find that a large emission reduction is unnecessary.[51]

Rather than relying entirely on a wait-and-see strategy, Singer has proposed some ideas for dealing with global warming. In addition to his aforementioned suggestion for a "no regrets" conservation policy, he has also supported a carbon sequestration plan involving ocean fertilization. While conventional approaches to carbon dioxide sequestration have called for tree planting on a massive scale, which would require vast areas of land and huge expenditures of money, fertilization of the oceans for the purpose of stimulating phytoplankton growth would speed up natural absorption of carbon dioxide into the ocean without, according to Singer, "depressing the economies of industrialized nations or limiting the economic growth options of developing nations." Furthermore, an increase in the population of phytoplankton, which constitutes the base of the oceanic food chain, would increase the populations of larger fish, leading to the development of new commercial fisheries in areas currently devoid of fish. Carbon dioxide from fossil fuel burning, Singer enthuses, would "thus becomes a natural resource for humanity rather than an imagined menace to global climate."[52]

GLOBAL WARMING BELIEVERS

Just as skeptics have criticized the methods and conclusions of climate scientists who warn of the dangers of global warming, environmentalists and believing scientists have questioned the honesty and motives of the skeptics. One of the more outspoken critics of the skeptics has been Ross Gelbspan, a Pulitzer Prize–winning journalist who wrote extensively about environmental issues during his career.

According to Gelbspan, complaints by skeptics that GCMs are inaccurate are hopelessly obsolete, as recent improvements in the computers have left little doubt that the models are correct.[53] He claims that a "dozen or so" skeptics have used this outdated argument to contradict the consensus view held by 2,500 of the world's top climate scientists and to keep discussions about global warming focused on whether there really is a problem, thereby preventing meaningful discussion about how to address the problem.[54] The notion that taking action to reduce greenhouse emissions will create economic chaos is also erroneous. Gelbspan has cited the research of numerous economists who have shown that the longer an energy transition from fossil fuels to renewable energy is delayed, the costlier the transition will be.[55] To avoid acting now, he claims, "could be to compound, incalculably, the costs of addressing climate change and its disruptions to civilization."[56] Other arguments by skeptics also fail under scrutiny. Estimates of agricultural benefits in a warmer world, for example, ignore "the fact that crops in the developing world would be decimated."[57]

Gelbspan reserves his harshest criticism for the perceived connection between skeptics and the fossil fuel industry, which he believes has used Balling, Michaels, Singer, Lindzen, and others to debunk global warming. In May 1995, while testifying in a St. Paul, Minnesota, courtroom about the environmental impact of burning coal, Michaels revealed under oath that he had received more than $165,000 in industry and private funding during early 1990s and that many of his publications have been financed by the coal industry. (Gelbspan has registered the same complaint about industry-funded research that Singer has made against IPCC research: that it is not necessarily peer reviewed and is often published without undergoing rigorous scientific scrutiny, as government-funded research is required to do.) Balling also admitted to having received hundreds of thousands of dollars in funding from the coal industry, which he revealed only after being questioned under oath. According to Gelbspan, Singer is the only prominent skeptic who has readily admitted to being funded by large oil companies. Singer defends himself by saying that his scientific opinions predate funding, but environmentalists have questioned what would happen if Singer came across information proving that fears of global warming are justified. Would he reveal the data, they ask, if it meant losing his industry funding?[58]

The ongoing global warming debate is capable of raising passions to a level that provokes extremist thought and behavior on both sides. The radical environmentalist group Earth Liberation Front (ELF) has been investigated for defacing sport-utility vehicles (SUVs), a protest likely stemming from anger over rising levels of greenhouse gas emissions.[59] Although the traditional role of a scientist is to conduct research and report data in an unbiased manner, the global warming debate has prompted some researchers to cross into the realm of personal values in their public comments. Skeptic Lindzen, for example, has said that the group behavior and dynamics of the environmental movement are similar to those of the Nazi movement in Weimar Germany, except for the anti-Semitism.[60]

THE POLITICIZATION OF GLOBAL WARMING

Skeptics and believers have each accused the other of "politicizing" global warming, and indeed many of their battles have been played out in the arena of global warming policy debates, both domestic and international. The debate began in earnest in 1989, the year after the formation of the IPCC and James Hansen's warning to the U.S. Senate Committee on Energy and Natural Resources that "the greenhouse effect has been detected and it is changing our climate now." In January, the British Meteorological Office announced that 1988 was the warmest year in the 127-year temperature record and that the six warmest years occurred in the 1980s, making it the warmest decade on record. Data collected by the researchers also independently confirmed that the average global temperature had risen by about 0.6 degree Celsius (1 degree Fahrenheit) since the beginning of the 20th century. Though admitting that these results were consistent with what would be expected from global warming, scientists involved with the study were reluctant to make an "unambiguous" connection between rising temperatures and greenhouse gas emissions.

In May 1989, U.S. president George H. W. Bush announced his support for the negotiation of a climate convention and invited delegates of the IPCC to a major workshop in Washington, D.C., to discuss the elements of such a convention. Several months later, at a summit in Malta, Bush proposed that the first negotiation session on a possible climate convention be held in the United States in the fall of 1990.

In the time between these two invitations, Richard Lindzen and Reginald Newell of MIT, along with Jerome Namias of Scripps Institution of Oceanography, sent a letter to Bush in support of a paper published by the George C. Marshall Institute, a conservative research group that provides technical advice on scientific issues. In the article, titled "Scientific Perspectives on the Greenhouse Problem," authors William A. Nierenberg, Robert Jastrow, and Frederick Seitz concluded that global warming could not be definitively attributed to fossil fuel emissions. Therefore, no government action should be taken to reduce greenhouse gases.[61] The following year, S. Fred Singer did his part to ensure that the government and citizens of the United States were aware of the skeptical point of view when he founded the Science and Environmental Policy Project (SEPP) to campaign against the idea that the Earth faced a serious threat from global warming and ozone depletion.

For the time being, however, momentum was on the side of the believers. In late 1989, delegates from 60 nations attended the first ministerial meeting on global warming, held in the Netherlands. Representatives of industrial nations agreed that carbon dioxide emissions should be stabilized as soon as possible. This resolve was given a boost early in 1990 with the release of the IPCC's First Assessment Report. Together, these two events provided the basis and impetus for formal negotiations toward an international agreement on climate change, culminating in the signing of the United Nations Framework Convention on Climate Change (UNFCCC) at the Earth Summit in Brazil in 1992.

United Nations Framework Convention on Climate Change

In June 1992, delegates from more than 160 nations met at the United Nations Conference on Environment and Development (UNCED; also known as the Earth Summit) in Rio de Janeiro, Brazil, to address global environmental problems. After a week of negotiations, summit participants adopted Agenda 21, an action plan that called for improving the quality of life on a global scale in the 21st century by using natural resources more efficiently, protecting global commons (those natural systems and cycles that provide the foundation for healthy ecosystems around the world), better managing human settlements, and reducing pollutants and chemical waste.

Negotiators also adopted the United Nations Framework Convention on Climate Change (UNFCCC), drafted the previous month in New York with the aim of stabilizing greenhouse gas concentrations in the atmosphere at levels that would prevent "dangerous interference with the climate system." The document was based on the core ideas that scientific uncertainty is not a valid excuse for avoiding precautionary action; that all nations have a common interest in confronting global warming, but specific responsibility may differ from one nation to the next; and that industrial nations have historically contributed the greatest amount of carbon dioxide to the atmosphere and are therefore responsible for taking the lead in addressing the problem of global warming.

The UNFCCC, which entered into force on March 21, 1994, committed industrial nations to a voluntary goal of returning carbon dioxide emissions to 1990 levels by 2000 and providing technical and financial assistance to developing and transitional nations. Signatory nations also agreed, among other things, to develop systems for keeping track of anthropogenic emissions and carbon sinks in their own countries; promote the application and transfer of technologies and processes that control emissions; promote sustainable management of greenhouse gases sinks and reservoirs; draft plans for coastal zone management; and participate in climate research and observation. However, the convention deferred agreement on the problem of precisely how developed nations were to limit their emissions.

Although the United States signed the UNFCCC, the tide of domestic global warming policy slowly began to turn in favor of the skeptics in the 1990s. Following the midterm elections of 1994, in which the House of Representatives and Senate saw an influx of a large number of conservative politicians, Congress voted to abolish the Office of Technology Assessment (OTA), a nonpartisan body of research specialists established years ago to screen the science on which policy makers depend.[62] The following year the Republican-led Congress cut 80 percent of the funding for an energy policy, enacted by the Bush administration in 1992 with bipartisan congressional support, that included $50 million for research and commercialization of wind energy. In May 1996, House Science Committee chairman Robert Walker successfully recommended cutting funding for NASA's Mission to Planet Earth, a program designed to monitor changes in global climate and their effects, despite strong recommendations by the National Research Council (NRC) to continue full funding. In his decision, Walker cited denials of an imminent climate crisis published by the George C. Marshall Institute.[63] Meanwhile, Dana Rohrabacher, a congressman and the chair of the Subcommittee on Energy and Environment of the House Committee on Science, ended funding for several government research programs that monitored stresses on global environment. Along the way, he accused the world's leading climate scientists of "practicing religion rather than science."[64]

During this time, global warming believers were not without at least one powerful ally in Washington, D.C. Vice President Al Gore was well versed in environmental issues, having written the book *Earth in the Balance: Ecology and the Human Spirit* (1992), in which he wrote, "We must make the rescue of the environment the central organizing principle for civilization." He had been advocating national and international action to combat the threat of global warming since his days as a senator during the Ronald Reagan administration of the 1980s. It was Gore who urged President Clinton to formally adopt the UNFCCC's goal of reducing greenhouse gas emissions to 1990 levels by 2000. Gore also convinced Clinton that strong international action to combat global warming must be taken, resulting in the vice president's appearance at the 1997 United Nations climate summit in Kyoto, Japan.

THE KYOTO PROTOCOL

In 1995, signatories to the UNFCCC met at the First Annual Conference of the Parties to the UNFCCC (COP 1), in Berlin, Germany, where they concluded that voluntary commitments were inadequate to achieve their emissions goals. As a result, they adopted the Berlin Mandate, which called for a new round of negotiations aimed at developing legally binding commitments. The mandate also called for the exemption of 134 developing nations—including China, India, and Mexico—from whatever emissions caps were eventually put into place. Also in 1995, the IPCC released its Second Assessment Report, which declared that "the balance of evidence suggests that there is a discernible human influence on global climate." The following year delegates at the Second Annual Conference of the Parties to the UNFCCC (COP 2), in Geneva, Switzerland, endorsed the IPCC's findings. They also agreed that the report would be the basis for new legally binding commitments to be negotiated at the Third Annual Conference of the Parties to the UNFCCC (COP 3), in Kyoto, Japan, in 1997.

From December 1 to December 11, 1997, nearly 6,000 United Nations delegates from more than 160 countries met in Kyoto to negotiate what would become known as the Kyoto Protocol to the Framework Convention on Climate Change. Also in attendance were about 3,600 representatives of environmental groups and the fossil fuel industry.[65] Among the official representatives was the 60-member U.S. delegation led by Undersecretary of State Stuart Eizenstat. The role of the United States, already suspect because of its status as the biggest emitter of anthropogenic greenhouse gas emissions in the world, was placed in further doubt by Eizenstat's declaration that "We want an agreement, but we are not going [to Kyoto] for an agreement at any cost." From Washington, D.C., Gore seemed to support this resolve when he said, "We are perfectly prepared to walk away from an agreement that we don't think will work."[66] Gore's apparent backpedaling was likely based on a nonbinding resolution introduced in Senate by Nebraska Republican Chuck Hegel the previous July that urged the Clinton administration to reject any agreement that does not require developing nations to control their emissions. The resolution passed by a vote of 95 to 0.

Indeed, the question of who would pay for the protocol's stated goals—agreeing on legally binding targets and timetables for reducing greenhouse gas emissions, developing new energy sources, and making existing energy production more efficient—divided delegates into opposing groups. Developing countries pointed their fingers at wealthy nations like the United States and those in Europe, blaming them for causing most of the greenhouse emissions during the past century and insisting that they shoulder most of the burden of correcting the problem. Developed countries countered that cuts in emissions would hurt their economies, which would in turn affect the world economy.[67] In the end, Gore himself flew to Kyoto for the last days of the conference, where he told U.S. delegates to "show increased flexibility" in their negotiations and help end the impasse.[68]

The resulting Kyoto Protocol committed 38 industrialized and former Eastern bloc countries to reducing their combined greenhouse gas emissions by at least 5.2 percent from 1990 levels by 2008 to 2012. Developing countries, meanwhile, committed to continuing their efforts to monitor and address greenhouse emissions but were exempt from legally binding requirements to do so. The protocol includes several controversial loopholes designed to help developed nations reduce the costs and difficulty of achieving emissions targets. These include emission trading, which allows developed countries to increase their emission caps by purchasing part of another country's Kyoto allocation; joint implementation, which allows developed nations to earn credits when they jointly implement specific projects that reduce emissions; and the Clean Development Mechanism (CDM), which allows developed nations to earn credits for projects implemented with developing nations.[69] Wealthy countries are also allowed to subtract from their emissions calculations the amount of carbon dioxide removed from the air by trees and other carbon dioxide removed from the air by trees and other carbon sinks. Although some nations with high emissions would have refused to sign the protocol without the inclusion of these loopholes, discussion of how systems such as emission trading would operate were deferred to later meetings.[70]

Reactions to the Kyoto Protocol were, of course, mixed. Representatives of fossil fuel interests and skeptics reiterated their argument that efforts to achieve emissions caps would wreak havoc on the global economy. Environmentalists, generally pleased, saw the protocol as only the first small step in combating global warming, pointing out that even if all the signatory countries participated, it would only slow the increase of carbon dioxide emissions. A reversal of emissions trends would require more drastic measures in the future.

Some who agreed with the need to take action to combat global warming criticized the structure of the protocol itself, claiming that setting targets and timetables for controlling emissions was not the best way to control pollution. The trading of carbon permits from nation to nation, some warned, would rely on the weak force of international law, which lacks a system of enforcement and penalty sufficiently strong to prevent nations from withdrawing from the agreement if their allocation proved inconvenient.[71] In the long run, the loopholes provided by joint implementation, the Clean Development Mechanism, emission trading, and carbon sinks would undermine the protocol's effectiveness by allowing wealthy nations to achieve their goals without actually reducing their emissions.[72]

ECONOMIC CONSIDERATIONS

One of the primary arguments made by skeptics is that mitigating carbon dioxide emissions would have devastating impacts on the global economy. One industry organization, the National Association of Manufacturers, estimated that the Kyoto treaty would have cost the United States nearly 5 million jobs and $400 billion in lost GDP through 2012.[73] Others argue that the discipline of economics is at least as uncertain as that of climate science and that no one can be certain of the costs of curtailing greenhouse gas production. It could, many suggest, yield a net benefit by improving energy efficiency. Most agree that a workable strategy should fall somewhere between an immediate banning of all fossil fuel use, which would bring the industrial world to a standstill, and letting carbon dioxide emissions increase by doing nothing, which would force future generations to pay the massive costs of adapting to a warmer world, such as building dikes, moving coastal populations, and changing agricultural methods.

Many economists and scientists have pointed out that the cost of the Kyoto Protocol will ultimately depend on how it is implemented. Economist William Pizer of the Washington, D.C.–based Resources for the Future think tank and political scientist David G. Victor of the Council on Foreign Relations, both of whom have criticized the protocol, have argued that the treaty would be more viable if governments would allow companies to emit more carbon dioxide by paying a flat rate per ton if emissions reductions ever got too expensive. Although environmentalists are generally against such measures, Pizer and Victor argue that it would make nations more willing to participate and may clear the way for greater reductions. James Hansen of NASA's Goddard Institute for Space Studies suggested shifting the focus of mitigation efforts from carbon dioxide to other greenhouse gases such as methane. Among greenhouse gases, methane is third in quantity after water vapor and carbon dioxide, but it is, by volume, more potent, more threatening to local air quality, and less crucial to economic activity.[74]

Others suggest that the optimal amount of emissions reductions can be found by using Yale University economic professor William Nordhaus's Regional Integrated Climate–Economy (RICE) model, adapted from his earlier Dynamic Integrated Climate–Economy (DICE) model. The RICE model considers costs to the economic system stemming from both climate changes and emissions restrictions, and it takes into account the effects of global warming on agriculture, energy, forestry, water supply, sea level, and human health. It works by comparing the costs and benefits of business-as-usual scenarios (in which no action is taken to reduce emissions) to the costs and benefits of whatever emissions-reduction scenarios the researcher wishes to explore. In 1995, the model determined that the optimal carbon dioxide reduction amount would be 4 percent. Anything more would result in a net cost to society because the long-term environmental benefits would be outweighed by the economic costs; anything less than 4 percent would result in a net cost to society because more money would be spent in the future to adapt to higher temperatures than would be saved now by not cutting emissions.[75] Many environmentalists scoff at the idea of a 4 percent reduction, claiming that the climate crisis requires cuts closer to 70 percent in a very short time if humans wish to avoid the catastrophic effects of global warming.

This economic uncertainty is complicated by the fact that there seems to be a connection between economic effects and the stance that corporations take on the global warming issue. Insurance companies have sided with environmentalists in debates about climate change based on the fact that claims from violent weather were greater in 1998, for example, than in the entire decade of the 1980s. Insurers fear that increases in extreme weather, sea level rise, and other changes expected to accompany global warming will lead to even higher claims in the future.[76] On the other hand, as has been noted, fossil fuel companies fear that emissions caps will cost them business and revenue, and therefore tend to support the views of the skeptics. Between 1997 and 2000, however, companies such as British Petroleum (BP), DuPont, Royal Dutch/Shell, Ford Motor Company, DaimlerChrysler, Texaco, and General Motors withdrew from the Global Climate Coalition (GCC), which had been formed to oppose action to curb fossil fuel emissions. The GCC closed its doors in early 2002, claiming it had fulfilled its purpose of convincing the U.S. government to reject the Kyoto Protocol. Meanwhile, many businesses that had abandoned the GCC retreated from early hard-line skeptical stances and started exploring ways to take action on climate change. Automakers Honda and Toyota have already started selling hybrid cars, which combine gasoline and electric power to deliver lower tailpipe emissions and better gas mileage than traditional internal combustion engines. Most automobile manufacturers have announced plans to develop and market their own versions. Several automakers—including Ford, DaimlerChrysler and Honda—have conducted research into building cars powered by hydrogen, which does not release carbon dioxide when it is burned. BMW has even looked into burning hydrogen in conventional internal-combustion engines.[77] BP and Shell launched their own carbon dioxide emissions trading programs even before the implementation of the Kyoto Protocol was defined, and BP made its solar business into one of the world's largest and established a hydrogen division.[78] Still, BP is primarily an oil company, and its investment in alternative energy equals only about 1 percent of its $12.5 billion overall annual expenditures.[79]

Environmentalists insist that more must be done. The great hope of most global warming believers is to eventually replace fossil fuels with renewable power sources, such as solar, wind, geothermal, and biomass energy. The environmental group Greenpeace has said, "We are in a second world oil crisis. But in the 1970s the problem was a shortage of oil. This time round the problem is that we have too much." The only solution is choosing "a fundamentally new energy direction based on clean renewable energy, like wind or solar power." According to the environmental public-policy research organization Worldwatch Institute, "The only feasible alternative is a solar/hydrogen-based economy." The organization asserts that development of renewable energy technologies will require a monumental research effort conducted with the urgency of the Manhattan Project or the Apollo space program.[80]

GLOBAL WARMING POLICY AFTER KYOTO

The year following the Kyoto meeting was the warmest year since meteorologists began direct measurements of weather phenomena in the 1860s. Against this background, negotiators met at the Fourth Annual Conference of the Parties to the UNFCCC (COP 4), in Buenos Aires, Argentina, to discuss commitments by developing countries to fight global warming. Attendees agreed to an action plan and a timeline for finalizing the details of the Kyoto Protocol. Negotiations continued along similar lines at the Fifth Annual Conference of the Parties to the UNFCCC (COP 5), in Bonn, Germany, in 1999. The sixth conference, held in 2000 at The Hague, the Netherlands, collapsed when the European Union accused the United States, Japan, Canada, and Australia of seeking to use carbon sinks to avoid taking meaningful steps to cut domestic emissions. All efforts to find an acceptable compromise failed.

These difficulties in negotiating the Kyoto Protocol were a mere precursor to what was to come from the United States. Although Gore had added his signature to the treaty, it required a two-thirds majority vote in the U.S. Senate to become law. Doubting that the treaty had the necessary support, President Clinton never submitted it for a vote. Any hopes that the United States would become party to the protocol were quashed when George W. Bush took up residence in the White House following the 2000 election. On March 29, 2001, Bush announced that the United States would not ratify the Kyoto Protocol, claiming that the country's economy could sustain a reduction in carbon dioxide emissions. In a later speech, he referred to the Kyoto Protocol as "fatally flawed in fundamental ways" but committed the United States to "work within the U.N. framework and elsewhere to develop with our friends and allies and nations throughout the world an effective and science-based response to the issue of global warming."

The timing of Bush's decision to reject the protocol coincided with the release of the IPCC's Third Assessment Report, which cited "new and stronger evidence that most of the observed warming of the last 50 years is attributable to human activities." At current rates, the new study projected, temperature would increase by 1.4 to 5.8 degrees Celsius (2.5 to 10.4 degrees Fahrenheit) by 2100.

The assessment was immediately attacked by Richard Lindzen, who was one of Working Group I's numerous coauthors. Testifying before the Senate Environment and Public Works Committee, he claimed that "the vast majority" of writers of the full report had no role in the preparation of the summary for policy makers, the only portion of the assessment posted on the IPCC's web site and the only section that most people would read. Lindzen also alleged that IPCC coordinators pressured writers of the full report to describe the computer models upon which the conclusions were based more favorably than some authors wished.[81]

In June, the National Academy of Sciences (NAS) published a report commissioned by the White House in which a panel of scientists concluded that the IPCC policy summary provided an "admirable summary of research" and that human activity "very likely" had caused the increase in global temperatures since 1900. The NAS report conceded that uncertainties remain about the role of anthropogenic emissions because of gaps in knowledge about natural climate variations, but the committee also stated its belief, contrary to Lindzen's accusations, that the summary provided an accurate picture of the contents of the full report.

Bush responded by acknowledging the problem of global warming, but he emphasized the uncertainties mentioned in the report. He announced plans for programs to improve climate predictions and to create new technologies that would help monitor emissions and lead to the development of cleaner energy sources. However, he offered no plans to cut greenhouse emissions in the immediate future. Several months later, despite the conclusions of the NAS, Bush revealed his affinity with Lindzen when he accused IPCC chairman Robert Watson of intentionally slanting the Third Assessment Report's summary to support the theory of human-induced climate change. In April 2002, at the urging of the ExxonMobil oil company and the U.S. government, Watson was replaced as chairman of the IPCC by India's Rejendra Pachauri, the first non–atmospheric chemist to take the position.

The first international climate meeting after Bush's rejection of the Kyoto Protocol occurred in Bonn, Germany, in July 2001. Any illusion that successful negotiations depended on the involvement of the United States were shattered when representatives from 178 nations reached compromise agreements on such long-standing issues as emissions trading, carbon sinks, flexibility in meeting emissions targets, and special funds to help developing nations adapt to the impacts of climate change. Although the United States sent delegates to the meeting, they contributed nothing to the discussion. In November 2001, at the Seventh Annual Conference of the Parties to the

UNFCCC (COP 7), in Marrakech, Morocco, delegates finalized the legal text of the Kyoto Protocol's procedures and institutions. In late 2002, anticipating that the protocol would soon be ratified by the requisite 55 nations and come into effect in 2003, delegates met again, in New Delhi, India, to discuss a broad range of actions available to governments and civil society to address climate change.

On February 14, 2002, while the Kyoto Protocol was gaining strength on the international level, Bush announced plans to shift the domestic climate change policy toward voluntary action by industry. The Bush plan set voluntary goals for greenhouse gas emissions based not on emissions caps, as called for by the Kyoto Protocol, but on what he called greenhouse gas intensity, which refers to the ratio of emissions to U.S. gross domestic product. Industry groups applauded the move, claiming that the plan would allow room for economic growth. Environmentalists, however, lamented that such voluntary approaches to emissions reductions had failed for the past decade, and there was no reason to believe corporations would suddenly begin to comply. Environmentalists, and even White House Council of Environmental Quality chairman, James L. Connaughton, also pointed out that linking emission reductions to economic growth would actually cause emissions to increase by an additional 12 percent over the next several years, allowing the United States to produce at least 35 percent more greenhouse gases in 2010 than would be permitted under the Kyoto Protocol.[82]

Carbon dioxide emissions increased worldwide by more than 9 percent during the 1990s. Germany, the United Kingdom, and the former Eastern bloc nations actually managed to reduce emissions. Prominent among those whose emissions skyrocketed was China, a nation exempt from the Kyoto Protocol's greenhouse gas limits that is experiencing rapid economic growth. The United States, responsible for nearly one-quarter of global carbon output, saw emissions rise some 18 percent between 1990 and 2000.[83] Meanwhile, at the beginning of the 21st century, researchers reported that Alaskan glaciers were melting at an average rate of six feet, and in some cases hundreds of feet, per year, more than twice the rate previously assumed. Others noted that the phenomenon of coral bleaching was killing massive stands of coral in the Pacific Ocean and making surviving coral susceptible to disease.

Global warming believers look upon these and other revelations as pointing to a connection between carbon dioxide emissions and global warming. Skeptics maintain that global warming, if it is occurring at all, is at least partially the result of natural climate variation. Even if warming is caused by anthropogenic fossil fuel emissions, it will be largely beneficial to the human race. Even within the U.S. government, the debate remains ambiguous. In June 2002, the EPA concluded that greenhouse gas emissions produced by human activities were the primary cause of global warming. President Bush dismissed the report as a product of federal "bureaucracy" and later announced that computer climate models were not accurate enough to warrant agreement with international emission treaties such as the Kyoto Protocol. He promised, however, to develop a 10-year research plan to better understand climate change.

---

[1] Christianson, Gale E. *Greenhouse: The 200-year Story of Global Warming.* New York: Penguin, 1999, pp. 3–12.
[2] Lomborg, Bjorn. *The Skeptical Environmentalist: Measuring the Real State of the World.* Cambridge, England: Cambridge University Press, 2001, p. 260.
[3] Christianson, pp. 109–110.
[4] *Ibid.*, p. 115.
[5] Lomborg, p. 261.
[6] *Ibid.*, pp. 261–263.
[7] *Ibid.*, p. 262.
[8] Christianson, pp. 48–74.
[9] *Ibid.*, pp. 95–101.
[10] *Ibid.*, p. 142.
[11] *Ibid.*, pp. 149–160.
[12] Stein, Paul. *Global Warming: A Threat to Our Future.* Library of Future Weather and Climate Series. New York: Rosen, 2001, p. 22.
[13] National Assessment Synthesis Team. *Climate Change Impacts on the United States: The Potential Consequences of Climate Variability and Change.* Washington, D.C.: U.S. Global Change Research Program, 2000, p. 13.
[14] Stevens, William K. *The Change in the Weather: People, Weather, and the Science of Climate.* New York: Delta, 1999, p. 138.
[15] *Ibid.*, pp. 94–98.
[16] Nilsson, Annika. *Greenhouse Earth.* New York: John Wiley and Sons, 1992, p. 8.
[17] Stevens, pp. 145–146.
[18] Nilsson, p. 8.
[19] *Ibid.*, pp. 8–9.
[20] Stevens, p. 132.
[21] *Ibid.*, p. 291.
[22] *Ibid.*, p. 161.
[23] Worldwatch Institute. *State of the World 2002: A Worldwatch Institute Report on Progress Toward a Sustainable Society.* New York: W. W. Norton and Company, 2002, p. 26.
[24] Second IPCC Report cited in Houghton, John T., et al., eds. *Climate Change 2001: The Scientific Basis.* Cambridge, England: Cambridge University Press, 2001, p. 10.
[25] Stevens, pp. 161–168.
[26] Houghton, et al., eds. *Climate Change 2001*, pp. 2–4.
[27] *Ibid.*, pp. 5–10.

[28] *Ibid.*, pp. 12–16.
[29] National Assessment Synthesis Team, p. 108.
[30] Dryer, Nicole. "Antarctic Meltdown?" *Science World*, vol. 58, May 6, 2002, p. 4.
[31] Houghton, John. *Global Warming: The Complete Briefing.* Second edition. Cambridge, England: Cambridge University Press, 1997, pp. 108–115.
[32] Bryant, Dirk, et al. *Reefs at Risk: A Map-Based Indicator of Threats to the World's Coral Reefs.* Washington, D.C.: World Resources Institute, 1998, pp. 8, 9–10.
[33] Worldwatch Institute, pp. 9–10; and National Assessment Synthesis Team, p. 111.
[34] National Assessment Synthesis Team, p. 112.
[35] Christianson, pp. 223–224.
[36] National Assessment Synthesis Team, p. 100.
[37] Petit, Charles W. "The Great Drying: Blame It on Distant Oceans—and Perhaps on Global Warming." *U.S. News & World Report*, vol. 132, May 20, 2002, p. 54.
[38] National Assessment Synthesis Team, pp. 97–99.
[39] *Ibid.*, pp. 90–95.
[40] Stein, p. 38.
[41] Houghton, p. 131.
[42] Wohlforth, Charles. "Forest Killers." *Alaska*, vol. 68, March 2002, pp. 32–35.
[43] Houghton, p. 131.
[44] National Assessment Synthesis Team, pp. 102–107.
[45] Lomborg, p. 266.
[46] Singer, S. Fred. *Hot Talk, Cold Science: Global Warming's Unfinished Debate.* Oakland, Calif.: The Independent Institute, 1997, pp. 17–19.
[47] *Ibid.*, p. 2.
[48] *Ibid.*, p. 7.
[49] *Ibid.*, p. 20.
[50] *Ibid.*, p. 91.
[51] *Ibid.*, p. 64.
[52] *Ibid.*, p. 21.
[53] Gelbspan, Ross. *The Heat Is On: The High Stakes Battle Over Earth's Threatened Climate.* Reading, Mass.: Addison-Wesley, 1997, p. 19.
[54] *Ibid.*, p. 40.
[55] *Ibid.*, p. 26.
[56] *Ibid.*, p. 31.
[57] *Ibid.*, p. 26.
[58] *Ibid.*, pp. 41–16
[59] Bacon, Lisa. "Rash of Vandalism in Richmond May Be Tied to Environment Group," *New York Times*, November 18, 2002, p. 15.
[60] Gelbspan, p. 54.
[61] Christianson, pp. 200–201.
[62] Gelbspan, pp. 63–64.
[63] *Ibid.*, pp. 3–4.
[64] *Ibid.*, p. 4.
[65] Christianson, p. 255.
[66] *Ibid.*, p. 255.

[67] Stein, p. 49.
[68] Leggett, Jeremy. *The Carbon War: Global Warming and the End of the Oil Era*. New York: Routledge, 1999, p. 308.
[69] Victor, David G. *The Collapse of the Kyoto Protocol and the Struggle to Slow Global Warming*. Princeton, N.J.: Princeton University Press, 2001, p. 4.
[70] *Ibid.*, p. 7.
[71] *Ibid.*, p. x.
[72] *Ibid.*, pp. 8–9.
[73] "President's Stance on Global Warming Wins Support." *Chemical Market Reporter*, vol. 261, February 25, 2002, p. 25.
[74] Musser, George. "Climate of Uncertainty." *Scientific American*, vol. 285, October 2001, pp. 14–15.
[75] Lomborg, pp. 305–307.
[76] Worldwatch Institute, p. 6.
[77] "Replacing Gas with a Gas: Hydrogen-Powered Cars." *The Economist*, vol. 360, July 21, 2001, p. 66.
[78] Scott, Alex. "BP Experiments with CO_2 Emissions Trading." *Chemical Week*, vol. 160, October 28, 1998, p. 42.
[79] Worldwatch Institute, p. 41.
[80] Lomborg, p. 258.
[81] "No Consensus on Climate." *Oil and Gas Journal*, vol. 99, August 27, 2001, p. 19.
[82] "President's Stance on Global Warming Wins Support." *Chemical Market Reporter*, vol. 261, February 25, 2002, p. 25.
[83] Worldwatch Institute, p. 5.

CHAPTER 2

THE LAW AND GLOBAL WARMING

Issues relevant to global warming are covered by both state and federal laws. Several U.S. states have passed their own specific legislation that commits them to reductions in greenhouse emissions that exceed those called for by the federal government. Because global warming is a problem that crosses national boundaries, most countries (with the notable exception of the United States) have been receptive to efforts by the United Nations to deal with carbon dioxide emissions at the international level.

INTERNATIONAL TREATIES

Much attention and commentary on the subject of climate change has focused on United Nations treaties intended to take an international, multilateral approach to the mitigation of ozone depletion and global warming. Additional information about these treaties can be found on the United Nations Treaty Series web site (http://untreaty.un.org).

MONTREAL PROTOCOL ON SUBSTANCES THAT DEPLETE THE OZONE LAYER, WITH ANNEXES (CONCLUDED AT MONTREAL ON SEPTEMBER 16, 1987; ENTERED INTO FORCE JANUARY 1, 1989)

Recognition of the potentially harmful effects of depletion of the ozone layer led the international community to reinforce the framework laid out in the Vienna Convention for the Protection of the Ozone Layer in 1985. Signatories to the Montreal Protocol were obliged to adopt legislation and policies aimed at reducing activities likely to have an adverse effect on the ozone layer. Among these policies were a gradual phase-out of the various categories of controlled ozone depleting substances (ODS) and a banning of exports and imports of controlled substances from and to nonsignatory nations. Amendments signed in London (1990), Copenhagen (1992), Montreal (1997), and Beijing (1999) strengthened the protocol by adding new chemicals to the list of controlled substances and converting transitional substances into controlled substances, among other provisions.

The United States signed the Montreal Protocol on September 16, 1987, and ratified it on April 21, 1988. The protocol is considered by many to provide a model for successful international, multilateral agreements on environmental and atmospheric issues.

United Nations Framework Convention on Climate Change (Concluded at New York on May 9, 1992; Entered into Force on March 21, 1994)

The United Nations Framework Convention on Climate Change (UNFCCC), signed at the 1992 Earth Summit in Brazil, was designed to stabilize greenhouse gas concentrations in the atmosphere at levels that would prevent excessive interference with the Earth's climate. Among the core ideas of the convention were the principles that scientific uncertainty is not a valid excuse for avoiding precautionary action; that nations have "common but differentiated responsibilities"; and that industrial nations have historically contributed the greatest amount of carbon dioxide to the atmosphere and are therefore responsible for taking the lead in addressing the problem of global warming.

Developed nations committed to a voluntary goal of returning emissions to 1990 levels by 2000 and providing technical and financial assistance to developing and transitional nations. Signatory nations also agreed to keep national inventories of anthropogenic emissions and sinks; promote the application and transfer of technologies and processes that control emissions; promote sustainable management of greenhouse gas sinks and reservoirs; draft plans for coastal zone management; and participate in climate research and observation.

Kyoto Protocol to the Framework Convention on Climate Change (Concluded at Kyoto on December 11, 1997; Not Yet Entered into Force)

In 1995, signatories to the UNFCCC concluded that voluntary commitments were inadequate to achieve their goals and adopted the Berlin Mandate, which called for a new round of negotiations aimed at developing legally binding commitments. These efforts culminated in the Kyoto Protocol, which was adopted at the third session of the Conference of the Parties to the UNFCCC, held at Kyoto, Japan, from December 1 to December 11, 1997.

The protocol maintains similar objectives to the UNFCCC, which includes "the stabilization of atmospheric concentrations of greenhouse gases at a level that would prevent dangerous anthropogenic interference with the climate system." Toward this end, industrial and former Eastern bloc countries are committed to reducing their combined greenhouse gas emissions by at least 5.2 percent from 1990 levels by 2008 to 2012. The protocol includes several mechanisms designed to help nations reduce the costs and difficulty of achieving emissions targets, including the trading of emissions permits, the use of carbon sinks, and earning of credits through joint implementation or the clean development mechanism. Developing countries commit to continuing efforts to monitor and address their greenhouse emissions.

In 1998, participating nations agreed to an action plan and a timeline for finalizing the details of the protocol. In 2000, negotiations at The Hague, the Netherlands, collapsed under the pressure of disagreements between the United States and the European Union. The United States, one of the original signatories to the protocol, never ratified it and formally withdrew from negotiations in March 2001.

In Bonn, Germany, in July 2001, 178 nations reached compromise agreements on such key issues as emissions trading, carbon sinks, flexibility in meeting emissions targets, and special funds to help developing nations adapt to the impacts of climate change. Later that year, delegates finalized the legal text of the Kyoto Protocol's procedures and institutions during a meeting in Marrakech, Morocco. In 2002, anticipating that the protocol would come into effect in 2003, delegates met in New Delhi, India, to discuss a broad range of actions available to governments and civil society to address climate change.

The protocol is not yet in force. It will enter into force 90 days after being ratified by at least 55 nations, including developed nations that represent at least 55 percent of that group's 1990 emissions.

U.S. Energy Policy

During 2002, President Bush and the 107th Congress sought to overhaul the energy policies of the United States for the first time since 1992. The Republican-led House of Representatives approved a sweeping energy bill that addressed some aspects of global warming but also included controversial provisions that called for opening the Arctic National Wildlife Reserve (ANWR) to oil drilling. In response, the Democrat-led Senate approved an alternate bill—the Energy Policy Act of 2002— that rejected the Arctic provisions, instead focusing on energy conservation and energy-efficient measures. When the 107th Congress adjourned in November 2002, no compromise had been reached.

The 108th Congress worked quickly to revive the energy debate. On April 11, 2003, the House of Representatives approved the National Energy Policy Act of 2003. The U.S. Senate began debating its own version in mid-May 2003, after which the House and Senate would negotiate a final bill to send to President Bush for his signature. The process was expected to last well into the summer of 2003.

Republican control and strict debating rules prevented extensive discussion about climate change in the House, but several provisions of the House-passed bill were related to emissions standards, including earmarking $1.6 billion to develop technologies that would reduce pollution from coal-fired power plants. Another measure called for doubling the amount of corn-based ethanol that would have to be added to the nation's gasoline supply by 2015. (Advocates claim that ethanol is a cleaner-burning, domestic alternative to foreign oil.)

The House bill also provided $18 billion in tax breaks. Although many of these were designed to benefit the oil and gas industries, some encouraged the development of solar and wind-generated power, as well as the purchase of energy-efficient homes. The House rejected a proposal to raise fuel economy standards for all vehicles to 30 miles per gallon by 2010, as well as President Bush's call for tax credits to promote the purchase of gas and electric hybrid cars.

Among the more controversial aspects of the House bill was a revival of the provision to drill for oil in ANWR. In March 2003, however, the Senate had voted to strip a similar provision from its budget resolution, and many senators, both Republican and Democrat, vowed that an energy bill containing ANWR leasing provisions would ultimately fail. The National Energy Policy Act of 2003 faced additional alterations in the Senate, where bipartisan blocs of senators promised to add regulations to control climate change, including amendments calling for mandatory reporting of greenhouse gases. An amendment cosponsored by Senators John McCain and Joseph Lieberman promoted a trading system that would allow firms that produce less fuel-efficient cars to buy credits from companies whose vehicle lines exceed the limit, effectively forcing manufacturers to pay more for the privilege of producing SUVs and other low-efficiency vehicles.

By the time it is finalized, the new energy policy will almost certainly include amendments to the National Climate Program Act, the Global Change Research Act of 1990, and the Energy Policy Act of 1992. These laws, current as of mid-2003, are summarized below.

NATIONAL CLIMATE PROGRAM ACT (1978)

Enacted in 1978 and amended by an act of Congress in 2000, the National Climate Program Act was passed to establish a coordinated and comprehensive national climate policy and program. See U.S. Code 15 U.S.C. 2901–2908 for the complete text of the act (Appendix E).

GLOBAL CHANGE RESEARCH ACT OF 1990

The Global Change Research Act of 1990 was enacted to amend the National Science and Technology Policy, Organization, and Priorities Act of 1976 (a bill that established a general science and technology policy for the United States) in order to provide for improved scientific understanding of the effect of changes in the Earth system on climate and human well-being.

Title I established the Committee on Earth and Environmental Sciences under the auspices of the Federal Coordinating Council on Science, Engineering, and Technology to increase the effectiveness and productivity of federal global change research efforts. Title I also established an interagency U. S. Global Change Research Program to improve understanding of global change.

Title II, titled the International Cooperation in Global Change Research Act of 1990, called for discussions with other nations on international agreements to coordinate global change research and an international research protocol for cooperation on the development of energy technologies that have minimal adverse effects on the environment. Title II also established the Office of Global Change Research Information to disseminate to foreign governments and their citizens, businesses, and institutions scientific research useful in preventing, mitigating, or adapting to the effects of global change.

Title III directs the secretary of commerce to conduct a study on the implications of growth and development on urban, suburban, and rural communities. Based on this study, the secretary is directed to assist state and local authorities in planning and managing growth and development while preserving community character. For the complete text of the act, see U.S. Code 15 U.S.C. 2921–2961 (Appendix E).

ENERGY POLICY ACT OF 1992

The Energy Policy Act of 1992 overhauled the policies of the United States regarding energy use, distribution, and research, with an emphasis on improving energy efficiency. Several titles continue to be important to the overall U.S. strategy of reducing greenhouse gas emissions. Important provisions of the act were reauthorized in the Energy Conservation Reauthorization Act of 1998. Relevant titles of the original act are summarized below:

Title I—Energy Efficiency: This title establishes energy efficiency standards, promotes electric utility energy management programs and dissemination of energy-saving information, and provides incentives to state and local authorities to promote energy efficiency.

Titles III, IV, V, and VI—Alternative Fuels and Vehicles: These titles provide monetary incentives, establish federal requirements, and support the research, design, and development of fuels and vehicles that can reduce oil use and, in some cases, carbon emissions as well.

Titles XII, XIX, XXI, and XXII—Renewable Energy, Revenue Provisions, Energy and Environment, and Energy and Economic Growth: These titles promote increased research, development, production, and use of renewable energy sources and more energy-efficient technologies.

Titles XVI—Global Climate Change: This title provides for the collection, analysis, and reporting of information pertaining to global climate change, including a voluntary reporting program to recognize electric utility and industry efforts to reduce greenhouse gas emissions.

Title XXIV—Hydroelectric Facilities: This title facilitates efforts to increase the efficiency and electric power production of existing federal and nonfederal hydroelectric facilities.

Title XXVIII—Nuclear Plant Licensing: This title streamlines licensing for nuclear plants.

See Appendix B for relevant excerpts from the legislative summary of the Energy Policy Act of 1992. The complete text of federal energy policy is available in the U.S. Code 42 U.S.C. 13201–13556. See 42 U.S.C. 13381–13388 for the text of the legislation enacted by Title XVI of the act (Appendix E).

U.S. Policies and Measures to Reduce Greenhouse Gas Emissions

The U.S. government is currently pursuing a broad range of strategies to reduce net emissions of greenhouse gases, all of which are conducted in cooperation with private-sector parties. This section summarizes several existing federal climate change programs, divided into the industry sectors they affect. (See Appendix D for a more complete list of programs.)

Energy: Residential and Commercial

Residential and commercial buildings account for approximately 35 percent of U.S. carbon dioxide emissions from energy use, mostly from lighting, heating, cooling, and operating appliances. Following are descriptions of some of the key policies and measures designed to promote investment in energy-efficient products, technologies, and better management practices.

Energy Star for the Commercial Market

This program focuses on promoting high-efficiency buildings and providing business leaders with the information they need to undertake effective building improvement projects.

Commercial Buildings Integration

This program seeks to integrate energy-saving initiatives into all facets of building construction and renovation, including design, operation and maintenance, indoor environment, and control and diagnostics of heating, ventilation, air conditioning, lighting, and other building systems.

Energy Star for the Residential Market

This program provides guidance for homeowners on designing efficiency into kitchen, additions, and whole-home improvement projects. It works with major retailers and other organizations to help educate the public on energy-saving practices and offers a web-based audit tool and a home energy benchmark tool to help the homeowner implement projects and monitor progress.

Community Energy Program: Rebuild America

This program helps communities, towns, and cities save energy, create jobs, promote growth, and protect the environment through improved energy efficiency and sustainable building design and operation.

Residential Building Integration: Building America

This program works with industry to jointly fund, develop, demonstrate, and construct housing that integrates energy-efficiency technologies and practices.

Energy Star-Labeled Products

The strategy of this program is to work in partnership with utilities and appliance manufacturers to identify and label energy-efficient products that are available to consumers.

Building Equipment, Materials, and Tools

This program conducts research and development on energy-efficient building components and design tools, and issues standards and test procedures for a variety of appliances and equipment. Examples of components that can be used to increase the energy efficiency of buildings include innovative lighting, advanced air-conditioning and refrigeration, fuel cells, and advanced windows, coatings, and insulation.

ENERGY: INDUSTRIAL

About 27 percent of carbon dioxide emissions in the United States results from industrial activities. The primary source of these emissions is the burning of carbon-based fuels, either on site in manufacturing plants or through the purchase of generated electricity. Many manufacturing processes use more energy than is necessary. The following programs help to improve industrial productivity by lowering energy costs, providing innovative manufacturing methods, and reducing waste and emissions.

Industries of the Future

This program works in partnership with the most energy-intensive industries in the United States to enhance their long-term competitiveness and to accelerate research, development, and deployment of technologies that increase energy and resource efficiency. Key elements of the strategy include an industry-driven report outlining each industry's vision for the future, and promotion of renewable energy sources (wind power, geothermal energy, hydropower, bioenergy, and hydrogen fuels) and traditional nonemitting sources, such as nuclear energy.

Best Practices Program

This program offers industry the tools to improve plant energy efficiency and environmental performance, and increase productivity. Selected best-of-class large demonstration plants are showcased across the country, while other program activities encourage the replication of those best practices in other large plants.

Energy Star for Industry

This initiative enables industrial companies to evaluate and reduce their energy use in a cost-effective manner. Through established energy performance benchmarks, strategies for improving energy performance, technical assistance, and recognition for accomplishing reductions in energy, the partnership contributes to a reduction in energy use in the U.S. industrial sector.

ENERGY: SUPPLY

Electricity generation is responsible for about 41 percent of carbon dioxide emissions in the United States. Federal programs promote greenhouse gas reductions through the development of cleaner, more efficient technologies (including renewable energy) for electricity generation and transmission.

Renewable Energy Commercialization

This program consists of several programs to develop clean, competitive renewable energy technologies, including wind, solar, geothermal, and biomass. The program also works to achieve tax incentives for renewable energy production and use.

Climate Challenge

This program is a voluntary joint effort of the electric utility industry and the Department of Energy to reduce, avoid, or sequester greenhouse gases.

High-Temperature Superconductivity

High-temperature superconductors conduct electricity with high efficiency when cooled to liquid nitrogen temperatures. This program supports industry-led projects to capitalize on recent breakthroughs in superconducting wire technology, aimed at developing such devices as advanced motors, power cables, and transformers. These technologies would allow more electricity to reach consumers with no increase in carbon dioxide emissions.

Hydrogen Program

This program's mission is to support the development of cost-competitive hydrogen technologies that will reduce the environmental impacts of energy use and increase the use of renewable energy.

Clean Energy Initiative

Through the Clean Energy Initiative, the EPA promotes a variety of technologies, practices, and policies with the goal of reducing greenhouse gas emissions associated with the energy supply sector. Strategies include expanding markets for renewable energy, working with state and local governments to develop policies that favor clean energy, and facilitating combined heat and power and other clean "distributed generation" technologies in targeted sectors. The initiative includes the Green Power Partnership and the Combined Heat and Power Partnership programs.

Nuclear Energy Plant Optimization

This DOE program is working to improve the efficiency and reliability of existing nuclear power plants. It aims to resolve issues related to plant aging and applies new technologies to improve plant reliability, availability, and productivity, while maintaining high levels of safety. The DOE also supports the Nuclear Energy Research Initiative (NERI), which funds small-scale research efforts on promising advanced nuclear energy system concepts, and the Generation IV Initiative, which is currently preparing a technology road map that will set forth a plan for large-scale research, development, and demonstration of promising advanced reactor concepts.

Carbon Sequestration

This DOE program develops the applied science and demonstrates new technologies for addressing cost-effective, ecologically sound management of carbon dioxide emissions through capture, reuse, and sequestration. Its goal is to make sequestration options available by 2015. The program's technical objectives include reducing the cost of carbon sequestration and capture from energy production activities; establishing the technical, environmental, and economic feasibility of carbon sequestration using a variety of storage sites and fossil energy systems; determining the environmental acceptability of large-scale carbon dioxide storage; and developing technologies that produce valuable commodities from carbon dioxide reuse.

TRANSPORTATION

Cars, trucks, buses, aircraft, and other components of the transportation system are responsible for about one-third of U.S. carbon dioxide emissions. Emissions from the transportation sector are increasing rapidly as Americans drive more and use less efficient sport-utility and other large vehicles. Following are some of the programs the federal government has implemented to promote the development of fuel-efficient vehicles, research options for producing cleaner fuels, and a reduction in the number of vehicle miles traveled.

FreedomCAR Research Partnership

This partnership with U.S. automobile manufacturers promotes the development of hydrogen as a primary fuel for cars and trucks. It focuses on the long-term research needed to develop hydrogen from domestic renewable sources and technologies that utilize hydrogen, such as fuel cells. The program also focuses on developing technologies that will enable mass production of affordable hydrogen-powered fuel cell vehicles and the hydrogen-supply infrastructure to support them. The FreedomCAR program includes the FreedomFUEL program, designed to help develop a national hydrogen fueling system by speeding up development of hydrogen fuel refining, storage, and delivery systems.

Innovative Vehicle Technologies and Alternative Fuels

The DOE funds research, development, and deployment of technologies that can significantly alter current trends in oil consumption. Among these programs are Vehicle Systems Research and Development (which funds research and development for advanced power-train-technology engines, hybrid-electric drive systems, advanced batteries, fuel cells, and lightweight materials for alternative fuels), Clean Cities (which works to deploy alternative-fuel vehicles and build supporting infrastructure, including community networks), and the Biofuels Program (which researches, develops, demonstrates, and facilitates the commercialization of biomass-based, environmentally sound, cost-competitive technologies to develop clean fuels for transportation).

EPA Voluntary Initiatives

The EPA supports a number of voluntary initiatives designed to reduce emissions of greenhouse gases from the transportation sector. Among the Commuter Options Programs are the Commuter Choice Leadership Initiative (increases commuter flexibility by expanding transportation mode options, using flexible scheduling, and increasing work location choices), Parking Cash-Out (offers employees the option to receive taxable income in lieu of free or subsidized parking), and Transit Check (offers nontaxable transit benefits of up to $100 monthly). Transit programs seek to reduce greenhouse gas emissions by carrying more people per gallon of fuel consumed than those driving alone in their automobiles. The Congestion Mitigation and Air Quality Improvement Program aims to reduce pollutants and provide financial aid to states to fund new transit services, bicycle and pedestrian improvements, alternative fuel projects, and traffic flow improvements likely to reduce greenhouse gases.

Ground Freight Transportation Initiative

This voluntary program is aimed at reducing emissions from the freight sector through the implementation of advanced management practices and efficient technologies.

Clean Automotive Technology

This program includes research activities and partnerships with the automotive industry to develop clean, fuel-efficient automotive technology. Among its goals are the transfer of hybrid engines and power-train components from passenger car applications to the more demanding requirements of sport-utility and urban delivery vehicle applications.

INDUSTRY (NON–CARBON DIOXIDE)

Although carbon dioxide accounts for the largest share of U.S. greenhouse gas emissions, non–carbon dioxide greenhouse gases have significantly higher global warming potentials. For example, over a 100-year period, methane is more than 20 times more effective than carbon dioxide at trapping heat in the atmosphere.

Natural Gas STAR

This partnership between industry and the federal government works to reduce methane emissions from natural gas systems by improving management practices and eliminating market barriers to the profitable collection and use of methane that otherwise would be released to the atmosphere.

Coalbed Methane Outreach Program

This program works to develop technologies that can eliminate the methane emissions from coal mine degasification systems. It is also addressing methane emissions in mine ventilation air.

Partnership with Aluminum Producers

This partnership works to reduce aluminum-industry emissions of greenhouse gases with extremely long atmospheric lifetimes, including hydrofluorocarbons (HFCs), perfluorocarbons (PFCs), and sulfur hexafluoride (SF_6).

Environmental Stewardship Initiative

Originally implemented to limit emissions of HFCs, PFCs, and SF_6 in three industrial sectors (semiconductor production, electric power distribution, and magnesium production), this program is being expanded to additional sectors pending an assessment of the availability of cost-effective emission reduction opportunities.

AGRICULTURE

The U.S. government maintains a broad portfolio of research and outreach programs aimed at enhancing the overall environmental performance of domestic agriculture, including reducing greenhouse gas emissions and increasing carbon sinks.

AgSTAR and RLEP

These programs focus on simultaneously reducing methane emissions and increasing productivity of U.S. farms.

Nutrient Management Tools

This program is designed to reduce nitrous oxide emissions by focusing on improving the efficiency of fertilizer use. A database provides information that enables farmers to develop nutrient management plans and detailed crop nutrient budgets and to assess the impact of management practices on nitrous oxide emissions.

Conservation Programs

Conservation programs help reduce greenhouse gas emissions and increase carbon sequestration in agricultural soils. Among these is the Conservation Reserve Program, which helps farm operators conserve and improve soil, water, air, and wildlife resources by removing environmentally sensitive land from agricultural production and keeping it under long-term, resource-conserving cover.

FORESTRY

The U.S. government supports efforts to sequester carbon in both forests and harvested wood products.

Forest Stewardship

The U.S. Department of Agriculture's Forest Stewardship Program and Stewardship Incentive Program provide technical and financial assistance to owners of nonindustrial private forests, which provide about 60 percent of the U.S. timber supply. The acceleration of tree planting on these lands helps meet resource needs and provides ancillary benefits such as soil conservation, water quality protection and improvement, recreation, and increased uptake and storage of carbon dioxide.

WASTE MANAGEMENT

Federal waste management programs work to reduce municipal solid waste and greenhouse gas emissions through energy savings, increased carbon sequestration, and avoided methane emissions from landfill gas—the largest contributor to U.S. anthropogenic methane emissions.

Climate and Waste Program

This program was introduced to encourage recycling and source reduction for the purpose of reducing greenhouse gas emissions. The EPA has implemented several targeted efforts within this program to achieve its goals.

WasteWise works to reduce waste through voluntary or negotiated agreements with product manufacturers. The Pay-As-You-Throw Initiative provides information and education on community-based programs that provide cost incentives for residential waste reduction. The EPA also conducts outreach, technical assistance, and research efforts on the links between climate change and waste management to complement these activities.

Stringent Landfill Rule

Implemented by the Clean Air Act of 1996, the New Source Performance Standards and Emissions Guidelines (Landfill Rule) require large landfills to capture and combust their landfill gas emissions.

Landfill Methane Outreach Program

This program encourages small landfills not regulated by the Landfill Rule to capture and use their landfill gas emissions. Capturing and using landfill gas reduces methane emissions directly and reduces carbon dioxide emissions indirectly through the utilization of landfill gas as a source of energy, thereby displacing the use of fossil fuels.

CROSS-SECTORAL

The federal government works to reduce greenhouse gas emissions from energy use in federal buildings and transportation fleets by requiring federal agencies to purchase products that use no more than one watt in standby mode; directing the heads of executive departments and agencies to take appropriate actions to conserve energy use at their facilities, review existing operating and administrative processes and conservation programs, and identify and implement ways to reduce energy use; requiring all federal agencies to take steps to cut greenhouse gas emissions from energy use in buildings by 30 percent below 1990 levels by 2010; directing federal agencies in Washington, D.C., to offer their employees up to $100 a month in transit and van pool benefits; and requiring federal agencies to implement strategies to reduce their fleets' annual petroleum consumption by 20 percent relative to 1990 consumption levels and to use alternative fuels most of the time.

Federal Energy Management Program

This program reduces energy use in federal buildings, facilities, and operations by advancing energy efficiency and water conservation, promoting the use of renewable energy, and managing utility choices of federal agencies. The program accomplishes its mission by leveraging both federal and private resources to provide federal agencies the technical and financial assistance they need to achieve their goals.

State and Local Climate Change Outreach Program

This EPA program provides technical and financial assistance to states and localities to conduct greenhouse gas inventories, to develop state and city action plans to reduce greenhouse gas emissions, to study the impacts of climate change, and to demonstrate innovative mitigation policies or outreach programs. Other projects include estimates of forest carbon storage for each state, a spreadsheet tool to facilitate state inventory updates, a software tool to examine the air quality benefits of greenhouse gas mitigation, a study of the health benefits of greenhouse gas mitigation, and a working group on voluntary state greenhouse gas registries.

STATE LEGISLATION

In recent years, some of the most innovative environmental and energy policies in the United States have originated on the state level rather than the federal level. States have considerable jurisdiction over electricity rates and generation, air pollution regulation, land use, waste management, agriculture, and transportation, all areas that have direct bearing on climate change. A number of states have drafted policies in these areas specifically aimed at reducing emissions of carbon dioxide, methane, and other greenhouse gases. This is an important step for the United States in general, since approximately half of the U.S. states would rank among the top 60 national emitters of greenhouse gases if they were counted as sovereign nations. The annual carbon dioxide emissions of Texas, for example, are higher than those of France, though Texas and France are similar in geographic size.

Commitment to reducing greenhouse gas emissions varies widely from state to state. Those that have enacted innovative climate change policies have generally done so with the support of both Democrats and Republicans, as well as with the blessing of industries eager to claim part of the credit for early reductions. Contrary to claims from the federal government that economic chaos is an inevitable side effect of emissions abatement, industry in states that have taken on the problem of global warming has not suffered undue stress from the costs of adjustment. Among the states that have instituted innovative emissions abatement programs are California, Massachusetts, New Jersey, and Oregon.

California

On July 22, 2002, California governor Gray Davis signed into law a bill mandating the California Air Resources Board (CARB) to "adopt regulations that achieve the maximum feasible and cost-effective reduction of greenhouse gas emissions" from passenger cars, trucks, and SUVs, starting in model year

2009. California, where motor vehicles account for 57 percent of all global warming pollution, thus became the first state in the nation to require automakers to reduce greenhouse gas emissions from the vehicles they produce. Although the bill does not supply specific instructions for how automakers are to achieve these reductions, it does prohibit banning SUVs and increasing gas and vehicle taxes. The onus is therefore on engineers at the major car factories to develop technology to meet the new standards.

Environmentalists considered the legislation a major victory in the fight against global warming. California is the fifth largest economy in the world; it accounts for 10 percent of all new car sales in the United States, and it has historically led the nation in auto regulation. As with unleaded gasoline, catalytic converters, and hybrid cars—all of which appeared first in California—activists hope the lower-emission cars will eventually become the standard for the entire nation (transportation accounts for 33 percent of all greenhouse gas emissions in the United States).

Automakers vowed to challenge the legislation, based on their belief that California overstepped its authority and intruded into the federal government's jurisdiction. The complete text of the law can be found on the Internet at http://www.leginfo.ca.gov/pub/01-02/bill/asm/ab_1051-1100/ab_1058_bill_20020501_amended_sen.html.

Massachusetts

In April 2001, Massachusetts governor Jane Swift issued emission control plans (ECPs) that established carbon dioxide caps for six of the state's oldest power plants, which collectively produce 40 percent of the state's electricity. Each plant is required to reduce its carbon dioxide emissions 10 percent below late 1990 levels by 2004–06. The plants have several options to attain compliance, including changing fuel or generating technologies, or trading carbon dioxide reduction credits with other plants in the state.

The ECPs are required by the state's tough clean-air regulations, under which all power plants must significantly reduce air pollutants. The regulations call for significant reductions in the emissions that cause or contribute to the formation of smog, acid rain, regional haze, mercury deposition, and global climate change. Each ECP outlines the steps facilities will take to reduce emissions of nitrogen oxide, sulfur dioxide, and carbon dioxide.

The six facilities named in the April 2001 mandate are Brayton Point Station in Somerset, Montaup Station in Somerset, Salem Harbor Station in Salem, Mount Tom Station in Holyoke, Canal Electric in Sandwich, and Mystic Station in Everett. The ECP approvals are available for review at the Massachusetts Department of Environmental Protection's web site at http://mass.gov/dep/bwp/daqc/daqcpubs.htm#ecp.

New Jersey

In 1998, then-governor of New Jersey Christine Whitman (later appointed chief of the EPA) issued the New Jersey Sustainability Greenhouse Gas (GHG) Action Plan by executive order. The plan called for a reduction of the state's greenhouse gas emissions by 3.5 percent below 1990 levels by 2005, putting New Jersey on schedule to meet the reduction goals pledged under the Kyoto Protocol by the end of the decade.

The plan identifies the major sources of greenhouse gases by source and sector in 1990. More than 80 percent of the greenhouse emissions in New Jersey result from the combustion of fossil fuels to produce energy for heating, cooling, electricity, and transportation. The GHG Action Plan identifies "no regrets" strategies designed to achieve the state's reduction goals. "No regrets" strategies include actions that are currently readily available, pay for themselves within the short term, and provide environmental benefits.

To meet its goals, New Jersey has implemented a comprehensive strategy that includes input from every relevant state agency, from agriculture to transportation. The recommended strategies are organized under five categories: energy conservation, innovative technologies, pollution prevention, waste management and landfill gas recycling, and natural resources and open space. Of particular note is the New Jersey Greenhouse Gas Registry, which facilitates crediting and trading of carbon dioxide emissions within the state. New Jersey also signed an agreement with the Netherlands that allows emission trading projects between the U.S. state and the European nation. Fact sheets and updates about the GHG Action Plan can be found on the Internet at http://www.state.nj.us/dep/dsr/gcc/gcc.htm.

Oregon

In 1997, the Oregon legislature gave the Energy Facility Siting Council authority to set carbon dioxide emissions standards for new energy facilities. State laws require that carbon dioxide emissions from any new power plant proposed for operation in Oregon must be at least 17 percent below those of the most efficient natural gas-fired plant now operating in the United States. This standard may be met either by developing cleaner, more efficient technologies or by purchasing carbon dioxide offsets through the Oregon Climate Trust, which has developed carbon dioxide mitigation projects such as solar rural electrification, geothermal heating, reforestation, and methane reuse from coal mines and sewage treatment plants. More information on Oregon's carbon dioxide standard can be found on the Internet at http://www.energy.state.or.us/siting/co2std.htm.

CHAPTER 3

CHRONOLOGY

This chapter presents a chronology of eras and dates significant to the study of global warming. Although it supplies relevant information on geological and historical events of the distant past, the chronology focuses primarily on the period following the emergence of global warming as a matter of public debate and international policy during the 1980s.

345 to 280 Million Years Ago

- During the Carboniferous period, a division of the Paleozoic era, the proliferation of life in the form of giant ferns, early amphibians, and insects begins to remove increasing amounts of carbon dioxide from the atmosphere. The decay of these organisms in large swamps inaugurates the formation of great stores of fossil fuels.

250 Million Years Ago

- Toward the end of the Permian period, also a division of the Paleozoic era, the Earth's landmasses merge into a single supercontinent known as Pangaea. Drastic climate changes lead to the extinction of 90 percent of all ocean species, 70 percent of reptile and amphibian families, and 30 percent of insect orders. The global temperature is 8 degrees Celsius (14.4 degrees Fahrenheit) warmer than it is today.

140 to 65 Million Years Ago

- During the Cretaceous period, a division of the Mesozoic era, the average global temperature is 5.5 to 11 degrees Celsius (10 to 20 degrees Fahrenheit) higher than today. This is the period of dinosaurs and alligators, which abruptly ends with another mass extinction.

65 to 55 Million Years Ago

- The Earth's atmospheric carbon dioxide level leaps to 3,000 to 3,500 part per million (ppm)—eight to ten times the current level—possibly due to methane hydrate releases or volcanic eruptions. The large-scale burial of the surplus carbon through swamps leads to the massive formation of coal.

50 Million Years Ago

- The Earth's climate begins a long, pronounced cooling trend that lasts into the modern era.

5 Million to 2 Million Years Ago

- During the Pliocene epoch, large carnivores and the earliest hominids (humanlike primates) appear. The global temperature is 2.5 degrees Celsius (4.5 degrees Fahrenheit) warmer than today.

2 Million Years Ago

- The disappearance of early hominids from the fossil record corresponds with the appearance of the first members of the genus *Homo*, marking the origin of the human line to which *Homo sapiens* belongs.

900,000 Years Ago

- A pronounced drop in global temperatures inaugurates the onset of a pattern of repeated ice ages that last about 100,000 years and that are interspersed with shorter, warmer interglacial periods of 8,000 to 40,000 years. This pattern of fluctuation remains in place to this day.

18,000 Years Ago

- The most recent ice age comes to a close, and global temperatures begin to rise.

12,000 Years Ago

- During the Younger Dryas event, the world plunges abruptly back into near-glacial conditions for about 200 years. Scientists are at a loss to explain why this occurred. Paleoclimatic evidence suggests that the event ended as suddenly as it began, with temperatures in Greenland rising in 5.5-degree-Celsius (10-degree-Fahrenheit) jumps in less than a decade.

10,000 Years Ago

- The Holocene interglacial period, the most recent of the warm intervals between ice ages, begins. Rising global temperatures coincide with the development of agriculture and the domestication of animals by humans, leading to a radical increase in our impact on the environment. The Holocene interglacial period continues to this day.

A.D. 900 to 1300

- The Medieval Warm Period (also known as the Little Climatic Optimum), an era of slightly warmer temperatures and increased rainfall in the Northern Hemisphere, facilitates the expansion of agriculture, thereby allowing Viking explorers to settle in previously glacier-bound areas of Greenland and North America.

1300 to 1900

- The Little Ice Age, a period during which average temperatures in the North Atlantic region drop by about 1.5 to 2 degrees Celsius (2.7 to 3.6 degrees Fahrenheit), leads to extensive glaciation and shorter agricultural growing seasons in the Northern Hemisphere.

1600s

- The Scientific Revolution prompts serious exploration of the material world, including the atmosphere. The invention of new measuring instruments—including the thermometer (temperature), barometer (air pressure), and hygrometer (humidity)—allows scientists to make accurate and extensive observations of weather and climate for the first time.

1700s

- The Industrial Revolution begins. The rapid growth of industry in Europe and North America leads to a massive increase in the burning of fossil fuels for energy. The relatively recent rise in the concentration of carbon dioxide in the atmosphere is usually dated to the beginning of the Industrial Revolution.

1824

- Jean-Baptiste-Joseph Fourier inaugurates investigation into the greenhouse effect when he publishes "General Remarks on the Temperature of the Terrestrial Globe and Planetary Spaces." In the article, the French mathematician and physicist describes the Earth's atmosphere as a giant glass dome that traps some of the heat reflected from the Earth back into the atmosphere, thus keeping the planet warm enough to sustain life.

1859

- *August 26:* The first oil well is drilled at Oil Creek, Pennsylvania. Producing 10 to 35 barrels per day, the well nearly doubles the world's oil production. Thousands of oil derricks soon spring up across northwestern Pennsylvania. The following year, the world's first oil refinery is built near the site of the first oil well.

1860

- Belgian inventor Étienne Lenoir develops the first internal combustion engine.

1861

- Irish engineer John Tyndall publishes a paper in which he describes water vapor, carbon dioxide, and ozone as the atmosphere's most important heat absorbers. He suggests that even small reductions in the amount of carbon dioxide in the air could lower global temperatures, but he fails to consider the opposite effect: that more carbon dioxide could lead to global warming.

1896

- Swedish chemist Svante August Arrhenius, who will later win the Nobel Prize for work unrelated to climate change, advances the theory that alternating ice ages and interglacial periods are caused by long-term fluctuations in the amount of carbon dioxide in the atmosphere. He is also the first to propose that carbon dioxide emissions from the combustion of fossil fuels might heighten the greenhouse effect and lead to global warming. Arrhenius speculates that such warming would prove beneficial to humans.

1903

- Henry Ford founds the Ford Motor Car Company in Detroit, Michigan. Five years later, mass production of the Model T dramatically increases use of the internal combustion engine and makes petroleum one of the most important fuel sources in the world.

1934

- British meteorologist Guy S. Callendar publishes the article "The Artificial Production of Carbon Dioxide and Its Influence on Temperature," in which he reveals that the Earth's temperature has risen nearly 0.6 degree Celsius (1 degree Fahrenheit) since 1880. He links this warming trend to carbon dioxide released by the combustion of fossil fuels and estimates an additional 1.1-degree-Celsius (2-degree-Fahrenheit) rise in average global temperatures over the following century. Like Arrhenius, Callendar believes such warming will benefit humans by enhancing plant growth and delaying the return of glaciers.

1940

- The warming trend that began in 1880 comes to an end, marking the beginning of a cooling trend that will last into the mid 1970s.

1950

- ***March 5:*** Jule Charney, John Von Neumann, and other scientists at the Meteorology Project in Princeton, New Jersey, launch the first computerized simulation of the weather. The project eventually revolutionizes the use of computerized weather predictions, but it also leads to the realization that the behavior of the atmosphere is not fully deterministic, making it impossible to precisely predict the development of weather patterns.

1955

- Chemist Charles Keeling begins measuring carbon dioxide levels in the atmosphere. Although scientists had previously assumed that carbon dioxide levels vary from place to place, Keeling's measurements reveal a universal concentration of 315 parts per million (ppm). By the late 1990s, similar measurements will reveal global carbon dioxide concentrations of more than 360 ppm, the highest levels of the past 420,000 years.

1957

- Roger Revelle and Hans E. Suess, scientists with the Scripps Institution of Oceanography, report for the first time that the oceans are not absorbing as much of the carbon dioxide emitted to the atmosphere as had been previously thought. The significant amounts of carbon dioxide in the atmosphere, they claim, could cause global temperatures to rise. They call fossil fuel emissions "a large-scale geophysical experiment" with the Earth's climate.

1967

- Japanese physicist Syukuro Manabe and his colleagues at the U.S. federal government's Geophysical Fluid Dynamics Laboratory (GFDL) use an early computer simulation of the climate to calculate that a doubling of preindustrial carbon dioxide levels will lead to a global average temperature increase of 2.2 degrees Celsius (4 degrees Fahrenheit).

1968

- Syukuro Manabe and co-researcher Richard Wetherald develop the first three-dimensional computer model of the entire Earth. The new simulation estimates that a doubling of carbon dioxide levels will be accompanied by an average temperature increase of 3 degrees Celsius (5.4 degrees Fahrenheit).

1974

- ***June:*** Chemists Sherwood Rowland and Mario Molina publish a paper in the journal *Nature* advancing the theory that human-made chlorofluorocarbons (CFCs) may destroy ozone molecules, leading to erosion of Earth's protective ozone layer.

1975

- Syukuro Manabe and Richard Wetherald publish the findings of their computer model experiments, and Charles Keeling releases updated measurements of atmospheric carbon dioxide concentrations that show that levels are increasing. Taken together, the two articles alert many scientists to the idea that human activities can affect global climate.
- Ocean chemist Wallace S. Broecker warns that the current global cooling phase will, "within a decade or so, give way to a pronounced warming induced by carbon dioxide." Basing his conclusions on the study of ice core samples collected in Greenland, Broecker notes that the Earth has never remained cool when atmospheric carbon dioxide levels are as high as the current level of 330 ppm, and that there is no record of an ice age occurring except when levels have dipped below 200 ppm.

1976

- Researchers James Hansen and Yuk Ling Yung publish a paper in the journal *Science* that identifies methane and nitrous oxide as greenhouse gases.

1977–78

- An unusually harsh winter seems to confirm recent theories that a new ice age may be imminent. Several scientists, such as climatologist Reid Bryson, had speculated that human-made pollutants were reducing the amount of sunlight reaching the Earth. As evidence, he pointed to the global cooling trend and glacier expansion that began in the 1940s. By the late 1970s, however, the cooling trend had ended, and global temperatures were again on the rise.

1979

- ***June:*** George Woodwell of the Woods Hole Biological Laboratory, Gordon MacDonald of Mitre Corporation, and David Keeling and Roger Revelle of the Scripps Institution of Oceanography issue a report to the President's Council on Environmental Quality, concluding that "man is setting in motion a series of events that seem certain to cause significant warming of world climate unless mitigating steps are taken immediately."
- ***October:*** A study by the National Academy of Sciences (NAS) panel on climate change, initiated by President Jimmy Carter's science adviser in 1976, warns that a doubling of atmospheric carbon dioxide could raise global temperatures by 1.5 to 4.5 degrees Celsius (2.7 to 8.1 degrees Fahrenheit). The panel warns that "a wait-and-see policy may mean waiting until it is too late" to avoid significant global warming.

1980

- Representatives of the United Nations Environment Programme (UNEP), the World Meteorological Organization (WMO), and the International Council of Scientific Unions (ICSU) meet in Villach, Austria, to analyze the scientific basis of claims that carbon dioxide emissions are contributing to global warming. Attendees conclude that climate change is a major environmental issue but recommend additional research before a management plan to control emissions is developed.
- ***January:*** Gus Speth, chairman of the Council on Environmental Quality, urges policy makers to include concerns about carbon dioxide emissions and climate change in U.S. and global energy policies.

1981

- James Hansen publishes a study showing that the Earth's temperature had risen until 1940, cooled from 1940 to 1975, and began rising again during the mid 1970s. The net temperature change has been a warming of 0.4 degree Celsius (0.7 degree Fahrenheit). Hansen also predicts warming in the 21st century of "almost unprecedented magnitude" that could melt the West Antarctic ice sheet and cause the global sea level to rise by 4.5 to 6 meters (15 to 20 feet).

1983

- The U.S. Environmental Protection Agency (EPA) publishes "Can We Delay a Greenhouse Warming?" The report states that global warming will result in conditions in which "agricultural conditions will be significantly altered, environmental and economic systems potentially disrupted, and political institutions stressed." Despite this warning, the report concludes that a ban on coal use, the only way to effectively slow the rate of global warming, would be politically and economically unfeasible. In the same year, the U.S. National Academy of Sciences (NAS) publishes the report "Changing Climate," which finds that the accumulation of greenhouse gases in the Earth's atmosphere may eventually raise global temperatures by 1.5 to 4.5 degrees Celsius (2.7 to 8.1 degrees Fahrenheit). The report also states that current evidence is not enough to warrant changes in energy policy.

1985

- Scientists working on the British Antarctic Survey discover that January ozone levels over the Antarctic have dropped 10 percent below those of the previous year, causing a hole to appear in the Earth's ozone layer.
- ***October:*** A conference sponsored by the United Nations Environment Program (UNEP), the World Meteorological Organization (WMO), and the International Council of Scientific Unions (ICSU) convenes again in Villach, Austria, bringing together scientists from 29 nations. The conference report warns that "some warming . . . now appears inevitable" regardless of future actions and recommends consideration of a global treaty to address climatic change.

1986

- ***January:*** The World Meteorological Organization (WMO) and the National Aeronautics and Space Administration (NASA) issue a three-volume report on atmospheric ozone.
- ***June:*** At Senate hearings, James Hansen of NASA's Goddard Institute of Space Studies predicts a warming of 1.7 degrees Celsius (3 degrees Fahrenheit) over 30 years if current levels of carbon dioxide emissions continue. At the same hearing, NASA presents evidence of an ozone hole over Antarctica. Senator John Chafee asks the Environmental Protection Agency (EPA) and the Office of Technology Assessment to develop policy options for stabilizing greenhouse gas concentrations in the atmosphere. Congress responds by setting aside extra money in the EPA's budget to conduct climate change research.
- ***November:*** EPA administrator Lee Thomas says that the United States will propose an international phase-out of chlorofluorocarbons (CFCs).

Chronology

1987

- An ice core from Antarctica analyzed by French and Soviet scientists shows a close correlation between atmospheric carbon dioxide levels and temperature that dates back more than 100,000 years.
- ***January:*** Testifying before the Senate Environment and Public Works Committee, scientist Veerabhadran Ramanathan warns that past carbon dioxide emissions have already committed the planet to a warming of 0.7 to 2 degrees Celsius (1.3 to 3.6 degrees Fahrenheit). Wallace Broecker of the Lamont-Dougherty Geological Observatory claims that greenhouse warming could lead to large and unpredicted changes in global climate.
- ***September 16:*** The Montreal Protocol on Substances that Deplete the Ozone Layer is signed by 24 industrial nations, which commit to limiting production and use of chlorofluorocarbons (CFCs) to 50 percent of 1986 levels. The protocol is later strengthened to require industrial countries to phase out production of CFCs by 1996 and developing countries to do so by 2010.
- ***December:*** U.S. president Ronald Reagan and Soviet general secretary Mikhail Gorbachev commit to joint studies on global climate change.

1988

- ***January:*** President Reagan signs the Global Climate Protection Act, requiring the president to propose policy responses to Congress. Also this month, the United States and the Soviet Union set up a joint working group to develop response strategies to global warming.
- ***April:*** A report issued by the U.N. Environment Programme (UNEP) titled "Development Policies for Responding to Climate Change" claims that global warming will outpace the environment's ability to adapt to rising temperatures. Also in this month, the United States ratifies the Montreal Protocol.
- ***May:*** Delegates from 46 countries convene in Toronto, Ontario, at the International Conference on the Changing Atmosphere and suggest a reduction of global carbon dioxide emissions by 20 percent from 1988 levels by the year 2005. The prime minister of Norway, Gro Harlem Brundtland, calls for a global convention on the greenhouse effect.
- ***June 23:*** Testifying before a Senate committee, James Hanson, director of NASA's Goddard Institute of Space Studies, states that "global warming is now large enough that we can ascribe with a high degree of confidence a cause and effect relationship to the greenhouse effect." Hansen's testimony propels the issue of global warming into the public arena and into the forefront of international politics.

- ***August:*** U.S. vice president and presidential candidate George H. W. Bush states that "those who are worried about the greenhouse effect are ignoring the White House effect."
- ***November:*** Under the auspices of the World Meteorological Organization (WMO) and the United Nations Environment Programme (UNEP), the United Nations forms the Intergovernmental Panel on Climate Change (IPCC). Based in Geneva, Switzerland, this group of 2,500 prominent climate scientists is given the task of assessing the science on climate change, including causes, impacts, and possible responses.

1989

- ***January:*** The British Meteorological Office announces that its data indicates that 1988 was the warmest year in the 127-year temperature record and that the six warmest years occurred in the 1980s, making it the warmest decade on record. Furthermore, the average global temperature has risen by about 0.6 degree Celsius (1 degree Fahrenheit) since the beginning of the 20th century. Researcher David Parker, who helped collect data for the study, explains that while the temperature rise is "consistent with the 'greenhouse effect' caused by the rise in the levels of carbon dioxide and other man-made gases in the atmosphere, the earth's temperature fluctuates considerably due to natural causes, and no unambiguous connection can yet be made."
- ***March:*** British prime minister Margaret Thatcher leads a conference on ozone depletion for delegates from more than 100 countries. In the days leading up to the conference, European nations and the United States call for modifying the Montreal Protocol to commit signatories to a complete phase-out of CFCs.
- ***May:*** U.S. president George H.W. Bush and the British government announce their support for the negotiation of a climate convention. U.S. negotiators invite delegates of the IPCC to a major workshop in Washington, D.C., to discuss the elements of such a convention. Also in May in Helsinki, Finland, delegates from more than 80 nations at a UNEP meeting on the ozone layer sign the Helsinki Declaration, which calls for a revision of the Montreal Protocol that would phase out CFCs.
- ***September:*** The World Conference for Global Environment and Human Response Toward Sustainable Development, held in Japan, focuses on climate change and deforestation. Also during September, Richard Lindzen and Reginald Newell of the Massachusetts Institute of Technology (MIT), along with Jerome Namias of Scripps Institution of Oceanography, send a letter to President George H. W. Bush in support of the George C. Marshall Institute's "Scientific Perspectives on the

Greenhouse Problem." The report concludes that global warming cannot be definitively attributed to fossil fuel emissions and therefore no action should be taken to reduce greenhouse gases.

- *November:* Delegates from 60 nations attend the first ministerial meeting on global warming, held in the Netherlands. Industrial nations agree that carbon dioxide emissions should be stabilized as soon as possible.
- *December:* At the Malta Summit, President Bush proposes that the first negotiation session on a possible climate convention be held in the United States in the fall of 1990.

1990

- The Intergovernmental Panel on Climate Change (IPCC) releases its first assessment report, providing the basis for formal negotiations toward and international agreement on climate change. These negotiations culminate in the signing of the United Nations Framework Convention on Climate Change (UNFCCC) at Earth Summit in Brazil in 1992.
- S. Fred Singer founds the Science and Environmental Policy Project, (SEPP) think tank to campaign against the idea that the Earth faces a serious threat from global warming and ozone depletion.
- In Canada, the National Action Strategy on Global Warming is inaugurated to coordinate the efforts of federal, provincial, and territorial governments to address climate change. Also, the Energy Efficiency and Alternative Energy program is established to help Canadian homeowners and motorists reduce energy consumption.

1992

- The United Nations Conference on Environment and Development (also known as the Earth Summit) in Rio de Janeiro, Brazil, brings together delegates from most the world's nations, including 117 heads of state. Participants adopt Agenda 21, a voluminous blueprint for action that calls for improving the quality of life on a global scale by using natural resources more efficiently, protecting global commons, better managing human settlements, and reducing pollutants and chemical waste. Negotiators also adopt the United Nations Framework Convention on Climate Change (UNFCCC), whose aim is to "prevent dangerous interference with the climate system" and which establishes for industrial nations the nonbinding goal of returning carbon dioxide emissions to 1990 levels by the year 2000.

1993

- Canada develops the world's first hydrogen-powered fuel-cell transit bus, launched in Vancouver, British Columbia.

- *October:* U.S. president Bill Clinton and vice president Al Gore issue "The Climate Change Action Plan," which details the strategies the United States will use to conform to the UNFCCC.

1994

- Russian scientist Pavel Groisman and co-researchers find that the average annual extent of snow cover in the Northern Hemisphere has decreased by about 10 percent in the previous two decades.
- *March 21:* The UNFCCC, which was signed at the Earth Summit in Rio de Janeiro, Brazil, in 1992, enters into force. To date, it has been ratified by 181 countries.

1995

- The First Annual Conference of the Parties to the UNFCCC (COP 1) is held in Berlin, Germany. The resulting Berlin Mandate calls for the exemption of 134 developing nations—including China, India, and Mexico—from whatever limits on greenhouse gas emissions are decided upon in Kyoto, Japan, in 1997.
- The IPCC releases its second assessment report, which declares that "the balance of evidence suggests that there is a discernible human influence on global climate." This provides a new benchmark of international scientific consensus and becomes the basis for negotiating the 1997 Kyoto Protocol.

1996

- The Second Annual Conference of the Parties to the UNFCCC (COP 2) is held in Geneva, Switzerland. The report endorses the IPCC's contention that humans have made a "discernible" influence on global climate and that "projected change in climate will result in significant, often adverse, impacts on many ecological systems and socioeconomic sectors, including food supply and water resources and on human health."
- Biologist Camille Parmesan completes a rigorous five-year study of the range of the Edith's checkerspot butterfly. She finds that during the previous century, the butterfly's range shifted northward by about 160 kilometers (100 miles) and that populations have become more abundant at elevations above 2,400 meters (8,000 feet). These population shifts are, according to Parmesan, "exactly what is predicted from global warming scenarios."

1997

- *July:* Senator Chuck Hegel, a Nebraska Republican, introduces a non-binding resolution urging the Clinton administration to reject any agreement drafted at the upcoming Kyoto conference that does not require developing nations to control their emissions. The resolution passes by a vote of 95 to 0.

- ***December:*** The Third Annual Conference of the Parties to the UNFCCC (COP 3) is held in Kyoto, Japan. Delegates from more than 160 countries agree to the Kyoto Protocol, which strengthens the 1992 UNFCCC by mandating reductions of carbon dioxide emissions of 6 to 8 percent from 1990 levels by 2008 to 2012 for industrial countries. A controversial emissions-trading scheme and debates over the role of developing countries remain unresolved at the end of the conference.

1998

- The World Meteorological Organization (WMO) reports that 1998 was by far the warmest year on record and that seven of the 10 warmest years occurred in the 1990s (the other three having occurred since 1983).
- The ozone hole over Antarctica grows to 25 million square kilometers (9.7 million square miles), eclipsing the 1993 record of 3 million square kilometers (1.2 million square miles).
- A study by researchers at Switzerland's University of Zurich finds that glaciers in the European Alps have lost about 30 to 40 percent of their surface area and about half their volume since 1850.
- ***November:*** The Fourth Annual Conference of the Parties to the UNFCCC (COP 4) is held in Buenos Aires, Argentina. The international community discusses commitments by developing countries to fight global warming and agrees to an action plan to set guidelines and rules needed to implement the Kyoto Protocol. The United States adds its signature to the treaty but has yet to ratify it.

1999

- The Fifth Annual Conference of the Parties to the UNFCCC (COP 5) is held in Bonn, Germany, and includes discussions aimed at setting the rules for achieving the emissions targets set by the Kyoto Protocol.
- Researchers Michael E. Mann and Raymond S. Bradley of the University of Massachusetts and Malcolm K. Hughes of the University of Arizona release a study of temperatures dating back to A.D. 1000. They conclude that the 20th century was the warmest century and 1998 was the warmest year of the millennium.
- NASA scientists report that the southern half of the Greenland ice sheet has shrunk substantially between 1993 and 1998, losing about two cubic miles of ice each year.

2000

- The Sixth Annual Conference of the Parties to the UNFCCC (COP 6) is held in The Hague, the Netherlands. Delegates discuss a package of rules and guidelines for the Kyoto Protocol aimed at reducing emissions, but they fail to reach an agreement.

2001

- The IPCC releases its third assessment report, which cites "new and stronger evidence that most of the observed warming of the last 50 years is attributable to human activities." At current rates, the new study projects, temperature will increase by 1.4 to 5.8 degrees Celsius (2.5 to 10.4 degrees Fahrenheit) by 2100.
- ***March 29:*** President George W. Bush announces that the United States will not ratify the Kyoto Protocol, claiming that the country's economy cannot sustain a reduction in carbon dioxide emissions.
- ***May 2:*** Richard Lindzen, coauthor of portions of the IPCC's Third Assessment Report, testifies before the Senate Environment and Public Works Committee that "the vast majority" of writers of the full report had no role in the preparation of the summary for policy makers, the only portion of the assessment posted on the IPCC's web site and the only section that most people will read. Lindzen also alleges that IPCC coordinators pressured writers of the full report to describe the computer models upon which the conclusions were based more favorably than some authors wished.
- ***June:*** The National Academy of Sciences (NAS) publishes a report commissioned by the White House in which a panel of scientists concludes that the 2001 IPCC Third Assessment Report policy summary provides an "admirable summary of research" and that human activity "very likely" has caused the increase in global temperatures since 1900. The report concedes that uncertainties remain about the role of anthropogenic emissions because of gaps in knowledge about natural climate variations. President Bush acknowledges the problem of global warming but emphasizes the uncertainties mentioned in the report. In response, he announces plans for programs to improve climate predictions and to create new technologies that will help monitor emissions and lead to the development of cleaner energy sources. However, he offers no plans to cut greenhouse emissions in the immediate future.
- ***June 11:*** President George W. Bush delivers a speech at the White House in which he famously refers to the Kyoto Protocol as "fatally flawed in fundamental ways." He commits the United States to "work within the U.N. framework and elsewhere to develop with our friends

and allies and nations throughout the world an effective and science-based response to the issue of global warming."

- ***July 27:*** Galvanized by U.S. rejection of the agreement, representatives from 178 nations meet in Bonn, Germany, and finalize many of the Kyoto Protocol's key rules. Negotiators from the United States stand by and watch.
- ***October 29–November 9:*** The Seventh Annual Conference of the Parties to the UNFCCC (COP 7) is held in Marrakech, Morocco, where delegates finalize the legal text of the Kyoto Protocol's procedures and institutions.

2002

- ***January:*** A NASA study shows that the rate at which greenhouse gases enter the Earth's atmosphere has slowed during the past 20 years. Part of the decrease may be attributed to the 1987 Montreal Protocol, which phased out the use of such greenhouse gases as CFC's. The reasons behind reductions in the rate at which methane, carbon dioxide, and soot enter the atmosphere remain unclear.
- ***February 14:*** President George W. Bush announces plans to shift U.S. climate change policy, rejecting commitments made by the nation at the 1992 Earth Summit to reduce carbon dioxide emissions. Bush's plan calls for voluntary action by industry; this contradicts the 1997 Kyoto Protocol, which mandates reduction of greenhouse gases by industrial nations and which Bush rejected in 2001.
- ***February 27:*** The Bush administration announces its opposition to a Democratic proposal that would require dramatic improvements in fuel efficiency in cars and trucks sold in the United States.
- ***March 19:*** Scientists with the British Antarctic Survey announce that the Antarctic Peninsula's Larsen B ice shelf, 650 feet thick and with a surface area of 1,250 square miles, is collapsing with "staggering" rapidity. The Antarctic Peninsula, the scientists say, has warmed by 2.5 degrees Celsius (4.5 degrees Fahrenheit) over the past 50 years, much faster than the global average.
- ***April:*** Robert Watson is replaced as chairman of the Intergovernmental Panel on Climate Change (IPCC) by Rejendra Pachauri, the first non–atmospheric chemist to take the position. Watson's stance that action must be taken to combat global warming raised the ire of the U.S. government, which, under the influence of the ExxonMobil oil company, threatened to drop funding for the IPCC if Watson retained his chairmanship.
- ***April 18:*** The U.S. Senate votes to block oil and gas drilling in Alaska's Arctic National Wildlife Refuge, handing the Bush administration a key defeat in its federal energy plan.
- ***May:*** A 2,418-square-mile iceberg—nearly as large as Chesapeake Bay—breaks away from Antarctica.

- ***June 5:*** The U.S. Environmental Protection Agency (EPA) releases a report concluding that greenhouse gas emissions produced by human activities are the primary cause of global warming. President Bush dismisses the report as a product of federal "bureaucracy." EPA chief Christine Whitman claims that she did not know of the report until news organizations reported it, leading environmentalists to accuse her of being an "absentee landlord" at the EPA.
- ***June 20:*** A report published in the journal *Science* links global warming with a host of epidemics in plants and animals. Although focusing largely on wild plants and animals, the report implies that long-term warming could accelerate disease in humans as well.
- ***July:*** Attorneys general from 11 states—Alaska, California, Connecticut, Maine, Maryland, Massachusetts, New Hampshire, New Jersey, New York, Rhode Island, and Vermont—send a letter to President Bush, urging him to reconsider his stance on global warming and take action.
- ***July 11:*** The Bush administration announces that computer climate models are not accurate enough to warrant agreement with international emission treaties but promises to develop a 10-year research plan to better understand climate change. James L. Connaughton, chairman of the White House Council of Environmental Quality, acknowledges that "greenhouse gases will increase under our approach" but maintains that costly, near-term measures to reduce emissions are not justified by the current level of scientific certainty.
- ***July 19:*** A study by a team of researchers from the University of Alaska published in *Science* finds that Alaskan glaciers are melting at an average rate of six feet, and in some cases hundreds of feet, per year. This is more than twice the rate previously assumed, and it had accelerated during the seven or eight years previous to the study.
- ***July 22:*** California governor Gray Davis makes that state the first to fight global warming by requiring reduced tailpipe emissions of greenhouse gases. The law mandates that the greenhouse emissions of all passenger vehicles sold in the state must be reduced to the "maximum" economically feasible extent by 2009. The auto industry promises to challenge the new law in court.
- ***August:*** A study involving more than 5,000 scientists in 55 countries shows that pollution, over-harvesting, and climate change are destroying coral reefs around the world. The study finds that the phenomenon of coral bleaching, which occurs when rising temperatures cause the algae that populate and build coral to die off, has killed massive stands of coral in the Pacific Ocean and made surviving coral susceptible to disease.
- ***August 8:*** In a report published in *Nature*, climatologist David Travis at the University of Wisconsin and colleagues at Pennsylvania State Uni-

versity show that the three-day grounding of air traffic after the September 11, 2001, terrorist attacks against the United States led to the absence of a thin blanket of cirrus clouds that forms from water vapor exiting jet engines. The loss of this cloud cover caused a 2-degree-Celsius (3.5-degree-Fahrenheit) increase in the difference between the highest day temperature and lowest night temperature over the United States, proving that human-made cirrus clouds reduce heating during the day by blocking sunlight, while preventing cooling during the night by keeping infrared energy from radiating into space.

- ***September:*** The World Summit on Sustainable Development, also known as the Johannesburg Summit or JOWSCO, is held in Johannesburg, South Africa. More than 100 heads of state gather to draft a plan to implement the environmental goals outlined at the 1992 Earth Summit. The resulting Johannesburg Implementation Plan sets an ambitious, nonbinding agenda to deal with environmental problems ranging from clean water and toxic chemicals to poverty and global warming.
- ***September 20:*** A massive avalanche buries a Russian village under 3 million tons of ice and mud from a collapsing glacier. More than 100 people are killed. Scientists warn that continued climate-change induced melting of glaciers could lead to an increase in similar disasters around the world.
- ***October:*** Evidence of coral bleaching is found in the Hawaiian Islands for the first time.
- ***October:*** The government of Tuvalu, a tiny Pacific island nation endangered by sea level rise and coral reef die-offs, considers suing the U.S. government or large U.S. corporations for emitting greenhouse gases into the atmosphere and causing global temperatures to rise.
- ***October 23–November 1:*** The Eighth Annual Conference of the Parties to the UNFCCC (COP 8) is held in New Delhi, India. Delegates, anticipating that the Kyoto Protocol will go into effect in early 2003, discuss a broad range of specific actions available to governments and civil societies for addressing climate change.
- ***December:*** Climate scientists at the NASA Goddard Institute for Space Studies announce that 2002 was the second-warmest year in recorded history (1998 being the warmest and 2001 the third warmest) and the 25th consecutive year of above-average temperatures. During 2002, global temperature averages set records for the months of January, March, and September, while floating sea-ice cover in the Arctic Ocean shrank to record low levels. Also, a government report reveals that U.S. emissions of carbon dioxide and other greenhouse gases decreased by 1.2 percent during 2002, the largest annual decline in more than a decade. Experts attribute the decline to an economic slowdown and an unseasonably warm winter that sharply reduced demand for fossil fuels. Despite the decrease,

overall emissions were still 11.9 percent above the 1990 level, and the United States remained the single largest contributor to emissions linked to global warming.

2003

- ***January 2:*** Two studies published in the scientific journal *Nature* show that gradual warming over the past 100 years has caused animals to shift northward an average of nearly 6.5 kilometers (4 miles) per decade and has forced annual spring activities such as flowering, egg hatching, and migration to occur an average of five days earlier.
- ***January 3:*** Researchers at Lawrence Livermore National Laboratory publish evidence that the troposphere has heightened during the last two decades. The scientists attribute this change to the combined warming of the troposphere by greenhouse gases and cooling of the stratosphere caused by ozone depletion.
- ***January 8:*** Senators John McCain, a Republican from Arizona, and Joseph I. Lieberman, a Democrat from Connecticut, present a bill that would require industries to reduce emissions of heat-trapping gases to 2000 levels by 2010 and to 1990 levels by 2016. During the hearing, Senator James M. Inhofe, the Oklahoma Republican who is chairman of the Environment and Public Works Committee, criticizes the bill, saying that the causes of global warming are still open to question. Senator Ron Wyden, a Democrat from Oregon, counters that the Bush administration is doing nothing about climate change and that nothing "short of flooding the National Mall" would persuade the administration that global warming was a serious problem.
- ***January 13:*** NASA launches the *Ice Cloud and Land Elevation Satellite (ICESat)*, designed to measure ice sheets covering Greenland and Antarctica to determine whether global sea levels are rising or falling.
- ***January 28:*** President Bush announces plans for the $1.2 billion FreedomFUEL program to help develop a national hydrogen fueling system. The program is designed to augment the already existing FreedomCAR program (in which the Energy Department and U.S. auto manufacturers are conducting joint research to develop commercial use of hydrogen-powered vehicles) by speeding up development of hydrogen fuel refining, storage, and delivery systems.
- ***February 12:*** Bush administration officials announce agreements with 13 major industrial sectors (including electrical utilities, petroleum, mining, steel, semiconductors, and automobiles) for voluntary controls on emissions of gases linked to global warming. Environmental critics note that many of the announced emissions targets are based on "intensity" (the

amount of such gases per unit of economic production) rather than the absolute volume emissions, which would amount to a 13 percent increase in greenhouse gas emissions by 2012. Many opponents of regulation are also critical, claiming that the voluntary policies are precursors to mandatory ceilings.

- ***February 14:*** A study published in the *Proceedings of the Royal Society of London* reports that rising temperatures have caused North American red squirrels in the Canadian Yukon to breed an average of 18 days earlier in the spring than they did a decade ago. Scientists find that the change has occurred on the genetic level, as squirrels that are genetically apt to breed earlier have more reproductive success under warmer conditions, thereby causing a steady increase in that trait among the general squirrel population.
- ***February 19:*** A study published in the *Journal of Mammalogy* shows that the pika, a small rabbitlike mammal, has disappeared from nearly 30 percent of the areas where it was common in the early 20th century (primarily the western United States and Canada). The survey shows that areas that lost pikas were on average drier, warmer, and at lower altitudes than areas where populations were unaffected. Scientists claim that the die-off may be an early signal of what alpine and subalpine ecosystems across the globe will face if temperatures continue to rise.
- ***February 20:*** Attorneys general from seven states—Connecticut, Maine, Massachusetts, New Jersey, New York, Rhode Island, and Washington—announce plans to sue the EPA to force the federal government to regulate carbon dioxide emissions from power plants. The lawsuit contends that the EPA is violating the Clean Air Act by failing to analyze the health and environmental impacts of power plant emissions as it is required to do every eight years.
- ***February 25:*** An expert scientific panel, convened by the National Academy of Sciences at the Bush administration's request, issues a report criticizing the administration's proposed research plan on the risks of global warming. The report claims that the proposal "lacks most of the elements of a strategic plan" and that its goals cannot be achieved without far more money than the White House has sought for climate research. James R. Mahoney, director of the federal Climate Change Science Program, says that he welcomes the comments and that the final plan, scheduled for release in April, would most likely reflect some of the suggestions.
- ***March 7:*** In an article published in the journal *Science,* two Argentine researchers report on aerial surveys they conducted in 2001 and 2002, which found that the collapse of Antarctica's Larsen A ice shelf in 1995 led to a drastic speedup of the seaward flow of five out of six nearby inland glaciers. According to polar ice experts, the phenomenon suggests that seas might rise as much as several meters in the next several centuries.

- ***April 8:*** A group of 13 researchers backed by the Union of Concerned Scientists (UCS) predicts that local temperatures in the Great Lakes region will rise by 3 to 7 degrees Celsius (5 to 12 degrees Fahrenheit) in winter and 3 to 12 degrees Celsius (5 to 20 degrees Fahrenheit) in summer by the end of the 21st century.
- ***April 17:*** Ford Motor Company backs away from a pledge, made in 2000, to increase the fuel economy of its sport utility vehicles by 25 percent by 2005. Company executives claim that a lack of expected technological advances and of tax credits for hybrid vehicles forced them to extend the deadline, possibly to the end of the decade.
- ***May 2:*** The EPA reports that the average fuel economy of cars and trucks in the United States fell to its lowest level in 22 years in the 2002 model year, contributing heavily to the nation's rising oil consumption and carbon dioxide emissions.
- ***May 6:*** A report by the European Environment Agency shows that the European Union's collective greenhouse gas emissions have risen for the second year in a row. Germany and Great Britain (the European Union's largest economies) cut emissions by more than the amount stipulated by the Kyoto Protocol, while France just reached its goal. However, the report reveals that Italy, Spain, the Netherlands, Belgium, Greece, Austria, Portugal, Finland, Ireland, and Denmark are all falling short of their targets.
- ***May 14:*** A study published in *The Proceedings of the National Academy of Sciences* indicates that soot in the atmosphere probably contributes twice as much to global warming as estimated two years ago by an international panel of climate scientists. Some scientists counter that too little is known about sooty pollutants to reach firm conclusions about its effect, if any, on global climate.
- ***June 18:*** Drafts of the EPA's comprehensive review of U.S. environmental problems reveal that White House officials deleted references to studies citing carbon dioxide emissions as a likely cause of global warming. Some EPA staff members say the resulting report "no longer accurately represents scientific consensus," while others maintain that it will help policy makers address global warming.
- ***July 24:*** The Bush administration announces a 10-year study to determine whether greenhouse gases have contributed to global warming. Environmentalists argue that these questions have already been answered and the plan is a deliberate attempt to delay addressing the problem.
- ***July 31:*** The Senate passes an energy bill identical to one it passed the previous year, clearing the way for negotiations with the House of Representatives for the creation of sweeping changes in U.S. energy policy. The Senate also agrees to a fall vote on John McCain and Joseph Lieberman's landmark Climate Stewardship Act.

CHAPTER 4

BIOGRAPHICAL LISTING

This chapter provides brief biographies of important scientists, policy makers, and scholars who have been involved with issues relevant to climate studies in general and global warming in particular. Many of these individuals are or have been affiliated with organizations detailed in Chapter 8. Additional information on these and other people can be found in the sources listed in the bibliography in Chapter 7.

Svante August Arrhenius, Swedish chemist who originated the theory that long-term fluctuations in atmospheric carbon dioxide levels influence the surface temperature of the Earth and cause alternating ice ages and interglacial periods. Born February 19, 1859, Arrhenius earned his doctorate at the Uppsala University in Sweden. In 1895, he became a professor at Högskola, a technical university in Stockholm, Sweden. The following year he wrote "On the Influence of Carbonic Acid in the Air upon the Temperature of the Ground," an essay in which he advanced his theory correlating atmospheric carbon dioxide fluctuations with the occurrence of ice ages. Arrhenius was also the first scientist to suggest that fossil fuel combustion could add carbon dioxide to the atmosphere and lead to global warming. He calculated, however, that it would take 3,000 years of such emissions for the amount of atmospheric carbon dioxide to double, and he believed that any resulting warming would benefit humanity and the planet. In 1903, Arrhenius won the Nobel Prize in chemistry for research unrelated to global warming.

Roger Bacon, 13th-century British scientist and mathematician who, contrary to the consensus at the time, believed that science should rely on experimentation and empirical evidence rather than blind devotion to the authority of ancient philosophers. In the field of meteorology, Bacon accepted Aristotle's basic model of the atmosphere, but he adjusted Ptolemy's theory of climatic zones to account for the fact that mountains and other topographic features could block or alter the circulation of air, thereby influencing the surrounding region's climate.

Robert C. Balling, Jr., director of the Office of Climatology and an associate professor of geography at Arizona State University. Balling, who has published many papers and several books on various aspects of climate change and the greenhouse effect, believes that global warming does not pose a danger to humans or to the planet.

Tim P. Barnett, research marine physicist at Scripps Institution of Oceanography in La Jolla, California. He received his Ph.D. in physical oceanography from Scripps in 1966 and then worked at Westinghouse Ocean Research Laboratory and the U.S. Naval Oceanographic Office before rejoining Scripps as a faculty member in 1971. His research has focused on climate prediction and global climate dynamics. In the late 1980s, Barnett, along with Michael E. Schlesinger of Oregon State University, conducted landmark research on detecting the greenhouse "fingerprint," which consisted of comparing the climate's actual behavior with observations of reliable greenhouse indicators, such as tropospheric temperature, atmospheric pressure at sea level, and sea-surface temperature. As a result of this work, Barnett was named a lead author of the Intergovernmental Panel on Climate Change (IPCC) Second Assessment Report's chapter on detecting the signal of human-made climate change.

Ofer Bar-Yosef, Harvard University professor of anthropology and curator of Paleolithic archaeology in the Peabody Museum of Natural History at Yale University. Bar-Yosef has theorized that humans in the Middle East first made the transition from hunting and gathering to agricultural societies because long-term reductions in rainfall following the ice age forced them to grow and store food to survive long, harsh dry seasons.

Jacob Bjerknes, Norwegian meteorologist and the son of Vilhelm Bjerknes. As a member of the Bergen School, Jacob conducted a pioneering study of the characteristics and behavior of cyclonic storms, which enabled meteorologists to radically increase the accuracy of their storm forecasts.

Vilhelm Bjerknes, Norwegian physicist whose research during the early decades of the 20th century revolutionized understanding of the behavior of the atmosphere in the Northern Hemisphere. During the 1890s, Vilhelm developed mathematical equations describing how factors such as varying pressure and density determine the movement and circulation of fluids. He later applied this "circulation theorem" to his quest to combine observation with theoretical knowledge of the atmosphere's dynamics in order to develop precise weather predictions. While working for the Bergen Museum in Norway during World War I, Vilhelm recruited a large number of coastal citizens to file weather observations with him several times per day. This collection of data, combined with the circulation theorem and other physical principles, was used by Vilhelm and other researchers (collectively known as the Bergen School) to develop an overall picture of how the Earth's atmosphere functions and to lay the groundwork for modern meteorology.

Bert Bolin, Swedish greenhouse scientist who served as the first chairman of the IPCC (1988–97). After graduating from Sweden's Uppsala University in the mid 1940s, Bolin helped Bergen School alumnus Carl-Gustaf Rossby found a center for meteorology in Europe, where he focused his research on atmospheric gases and the carbon cycle. Upon Rossby's death, Bolin took over as director of the center but continued to study atmospheric gases. Bolin also worked as a professor in meteorology at the University of Stockholm, Sweden, and has served as an adviser to the Swedish government on environmental issues.

Wallace S. Broecker, Newberry Professor of Earth and Environmental Sciences at Columbia University's Lamont-Doherty Earth Observatory. His research has focused on the role of the oceans in climate change. In 1975, Broecker warned that the then-current global cooling trend would, "within a decade or so, give way to a pronounced warming induced by carbon dioxide." At the time of this prediction, atmospheric carbon dioxide levels were measured at 330 parts per million (ppm) and rising. Using data gleaned from ice core samples collected in Greenland, Broecker concluded that previous ice ages had occurred only when carbon dioxide levels had fallen below 200 ppm.

Reid Bryson, climatologist at the University of Wisconsin who, during the 1970s, argued that the planet faced a long-term global cooling trend caused by the large amount of Sun-blocking anthropogenic pollutants in the atmosphere.

George H.W. Bush, 41st president of the United States. During the election campaign in 1988, Bush said, "Those who think we are powerless to do anything about the greenhouse effect forget about the White House effect; as president, I intend to do something about it." In 1989, public opinion pressures convinced him to become one of the many world leaders who were calling for an international treaty on climate change. His administration also implemented a series of domestic "no regrets" actions aimed at promoting emissions reductions, such as strengthening efficiency standards for appliances, capturing methane from landfills, and planting more trees to offset carbon dioxide emissions.

George W. Bush, 43rd president of the United States. One of his first actions as president was his announcement, under the pretext that the nation's economy could not sustain a reduction in carbon dioxide emissions, that the United States would not ratify the Kyoto Protocol. In 2002, Bush rejected commitments made by the United States at the 1992 Earth Summit to reduce carbon dioxide emissions, announcing plans for voluntary action by industry. At the same time, he declared opposition to a Democratic proposal that would require dramatic improvements in fuel efficiency in cars and trucks sold in the United States. In April, the U.S. Senate voted to block oil and gas drilling in Alaska's Arctic National Wildlife Refuge, handing Bush a key defeat in his federal energy plan. Bush later said he believed that computer climate models were not accurate enough to warrant agreement with international emission treaties, but he promised to develop a 10-year research plan to better understand climate change.

Guy S. Callendar, British meteorologist who worked as a coal engineer and steam technologist for the British Electrical and Allied Industries Research Association. Using data collected from 200 weather stations around the world, he discovered that the average global temperature had risen nearly 0.6 degree Celsius (1 degree Fahrenheit) between 1880 and 1934. Callendar published these findings in 1938 in the paper "The Artificial Production of Carbon Dioxide and Its Influence on Temperature," in which he linked the warming trend to fossil fuel combustion and the resulting rise in the amount of carbon dioxide in the atmosphere. Although he estimated that temperatures would rise an additional 1.1 degrees Celsius (2 degrees Fahrenheit) over the next 100 years, Callendar, like Svante August Arrhenius before him, believed such warming would benefit humans by extending the agricultural growing season and delaying the advent of another ice age.

Jule Charney, theoretical meteorologist who pioneered the use of computers for weather prediction. In 1948, he was appointed director of the Meteorology Project in Princeton, New Jersey, where he worked alongside John Von Neumann and Carl-Gustaf Rossby. Charney's research focused on the problem of converting the physical properties that govern the behavior of the atmosphere into mathematical formulas that could be fed into a computer. When Charney and his colleagues launched the first computerized simulation of the weather on March 5, 1950, they were at the forefront of a revolution in computerized weather predictions. At the same time, however, the research led to the realization that it would never be possible to precisely predict the behavior of the atmosphere because the development of weather patterns is not fully deterministic. Charney was later appointed the Alfred P. Sloan Professor of Meteorology at the Massachusetts Institute of Technology. In 1979, he and other researchers, working under the auspices of the National Research Council, concluded that a doubling of carbon dioxide, predicted to occur during the early decades of the 21st century, would raise the average global temperature of the Earth by 1.5 to 4.5 degrees Celsius (2.7 to 8.1 degrees Fahrenheit).

Bill Clinton, 42nd president of the United States. In 1993, the Clinton administration formally adopted the Earth Summit goal of reducing greenhouse gas emissions to 1990 levels by 2000, which had previously been voluntary. In order to accomplish this, Clinton expanded the Bush administration's "no regrets" program to reduce emissions. The plan ultimately failed, as emissions in the year 2000 were 13 percent higher than in 1990. In 1997, Vice President Al Gore signed the Kyoto Protocol, but Clinton later retreated from his previous support of the document, failing to submit it to Congress for ratification.

Joseph Farman, British chemist who worked for Britain's Natural Environmental Research Council. Between 1957 and 1983, while working at the British Antarctic Survey's Antarctic outpost at Halley Bay, he used a spectrophotometer to measure various aspects of the atmosphere. In September 1981, Farman detected a decline in the amount of atmospheric ozone over Antarctica. Although the levels returned to normal after several months, lower readings were again detected in September of 1982 and 1983, each year growing progressively worse. By 1984, ozone levels were 40 percent below average. In 1985, Farman published a paper in which he reported his observations and warned of the growing ozone hole.

Chris Folland, analyst at the British Meteorological Office's Hadley Center for Climate Prediction and Research. He and co-researcher David Parker collected and analyzed global ocean temperature data, finding an upward trend in global temperature. The results of the project were published in 1984. Folland was appointed chief author of the 1990 IPCC report's chapter on observed climate change.

Henry Ford, U.S. businessman who founded the Ford Motor Company. The mass production of automobiles by Ford resulted in the exponential, worldwide growth of the petroleum industry.

Jean-Baptiste-Joseph Fourier, French scientist who first described the greenhouse effect. Born in 1768 in Auxerre, France, Fourier gained a reputation in Paris as a brilliant mathematician. Fourier worked on equations to explain how heat spreads. The resulting "diffusion equation," finally published in 1822, influenced heat theory in the fields of physics, theoretical astronomy, and engineering. In 1824, Fourier published "General Remarks on the Temperature of the Terrestrial Globe and Planetary Spaces," in which he became the first scientist to describe the Earth's atmosphere as a giant glass dome that traps some of the solar energy that is reflected from the Earth, allowing the planet to remain warm enough to sustain life. Fourier died of a heart attack on May 16, 1830.

Dian J. Gaffen, research meteorologist at the Air Resources Laboratory of the National Oceanic and Atmospheric Administration (NOAA). In 1998 Gaffen, along with co-researcher Rebecca J. Ross, analyzed hourly temperature and humidity observations at 113 NOAA weather stations nationwide. They found that, between 1949 and 1995, the frequency of extremely hot, humid days in the United States had increased by about two days a year per decade. They also found that apparent temperature, the most stressful form of heat, was increasing faster than temperature alone because humidity was increasing faster, a result that would be expected from a warming climate that caused more water to evaporate from the Earth's surface. In a paper published in the journal *Nature*, the researchers concluded that "regardless of the root causes, if these climate trends continue they may pose a public health problem, particularly in light of the growing population of elderly people most vulnerable to heat-related sickness and mortality."

W. Lawrence Gates, climatologist for the Program for Climate Model Diagnosis and Intercomparison at the Lawrence Livermore National Laboratory in California. In 1995, he led the IPCC's attempt to evaluate the accuracy of computer models of the atmosphere. As a result, an entire chapter of the IPCC's Second Assessment Report was dedicated to the topic.

Al Gore, Jr., U.S. politician who, as a senator from Tennessee during the Reagan administration, began strongly advocating national and international action to combat the threat of global warming. In 1992, his book *Earth in the Balance: Ecology and the Human Spirit* was published, in which he wrote, "We must make the rescue of the environment the central organizing principle for civilization." That year, he was also elected vice president of the United States as part of the Clinton administration. At Gore's urging, Clinton formally adopted the goal, drafted at the 1992 Earth Summit in Brazil, of reducing greenhouse gas emissions to 1990 levels by 2000. Gore also convinced Clinton that strong international action to combat global warming must be taken, resulting in the vice president's appearance at the 1997 United Nations climate summit in Kyoto, Japan. Once there, he told U.S. delegates to "show increased negotiating flexibility," which helped end an impasse and allowed an international agreement to be drafted. Although Gore added his signature to the protocol, President Clinton never submitted it to Congress for ratification. In 2000, while running a campaign for president of the United States, Gore continued to advocate action to mitigate the effects of global warming.

Biographical Listing

Pavel Groisman, research scientist at the University Corporation for Atmospheric Research. In 1994, while working at the State Hydrological Institute in St. Petersburg, Russia, Groisman collaborated with National Climate Data Center scientists Thomas Karl and Richard Knight on a study of snow cover. They found that over the previous two decades, the average amount of ground covered by snow each year in the Northern Hemisphere had decreased by about 10 percent. Groisman and his colleagues also discovered a feedback loop in which reduced snow coverage, caused by warmer atmosphere, further accentuated the warming by allowing more trapped heat to be released from the ground.

James E. Hansen, climatologist, director of that National Aeronautic and Space Administration (NASA) Goddard Institute of Space Studies, and member of the National Academy of Sciences. He earned a doctorate in physics from the University of Iowa in 1967 and started working full time at Goddard in 1972. Hansen began focusing his research on greenhouse gases in 1975, using computer models of the atmosphere to study the greenhouse effect. In 1981, he and his colleagues used surface temperature data collected from 2,000 weather stations around the world to determine that a warming trend that had begun in the late 19th century and leveled off between the 1940s and the 1970s had resumed. In 1983, Hansen contributed to the U.S. Environmental Protection Agency's report "Can We Delay a Greenhouse Warming?" which stated that as a result of global warming, "agricultural conditions will be significantly altered, environmental and economic systems potentially disrupted, and political institutions stressed." While testifying before the U.S. Senate Committee on Energy and Natural Resources on June 23, 1988, Hansen stated that "the greenhouse effect has been detected and it is changing our climate now" and that he was "99 percent confident" that current temperatures represented a "real warming trend" as opposed to natural variability. In making these statements, Hansen propelled the issue of global warming into the public arena and into the forefront of both domestic and international politics. Faced with criticism from scientists who did not share his convictions about climate change, Hansen offered to bet anyone that one of the first three years of the 1990s would be the warmest on record (he won the bet in 1990), and he predicted correctly that the 1990s would be the warmest decade on record.

Phil Jones, scientist at the Climatic Research Unit at the University of East Anglia in England. During the early 1980s, he and colleague Tom M. L. Wigley, along with analysts at the British Meteorological Office, embarked on a project to make sense of global climate data by collecting every available land temperature record in the world, combining them, and removing as many distorting influences as possible. The resulting analysis, updated on a monthly basis, has been widely cited by climate experts around the world. Jones contributed to the 1995 IPCC scientific report, has written numerous research papers, and has co-edited two books on paleoclimatology. He is also a fellow of the Royal Meteorological Society.

Thomas Richard Karl, director of the National Climate Data Center (NCDC) in Asheville, North Carolina. After being hired as an analyst at the NCDC in 1980, Karl began a seven-year project to enter all the available U.S. weather data into a computer database to give researchers a tool to detect and analyze climatic trends. As a result of this work, Karl was appointed one of the three lead authors for the IPCC First Assessment Report's chapter on observed changes in climate; he later contributed to the 1995 and 2001 reports as well. During the 1990s, he and his colleagues collected global temperature records showing increasing temperatures around the world. In 1993, Karl analyzed these databases to detect changes in extreme weather patterns and found an increase in heavy precipitation. He calculated that there was a 90 to 95 percent chance that this change was a result of global warming. In 1998, the year he became director of the NCDC, Karl began issuing monthly updates on global temperature trends.

Charles Keeling, U.S. chemist who conducted pioneering work in the measurement of atmospheric carbon dioxide levels. Keeling earned a Ph.D. in chemistry from Northwestern University and in 1953 began a temporary position in California Institute of Technology's new geochemistry program. To research the question of whether carbon dioxide in the atmosphere was in balance with the carbon dioxide in the oceans, Keeling was forced to build a device capable of measuring carbon dioxide in parts per million (ppm). He based his design on an article, published in 1916, describing the manometer, an instrument designed to measure small amounts of various gases. In 1955, he collected samples along the West Coast of the United States, always coming up with a measurement of 315 ppm of carbon dioxide. The following year Keeling began working at the Scripps Institution of Oceanography in La Jolla, California, where he improved the accuracy of the manometer by a factor of 10. In 1958, one of the new instruments was placed on Mauna Loa, a 4,104-meter (13,680-foot) volcano on the island of Hawaii. The data collected over the following decades have shown a steady increase in the levels of carbon dioxide in the atmosphere, an upward trend known as the Keeling curve.

Richard Lindzen, Sloan Professor of Meteorology at the Massachusetts Institute of Technology (MIT) and a member of the National Academy of Sciences. Lindzen, a graduate of Harvard University, has refuted the contention that global warming poses a threat to the health of the planet and humans. Among his arguments against global warming are his beliefs that atmospheric water vapor may diminish the warming effect of greenhouse gases rather than amplify it (as many scientists believe); that the short-term climate records available to scientists cannot be used to adequately detect long-term patterns of climate change; that computer models are not accurate enough to predict future climate trends; and that the Earth's atmosphere, which tends toward stability, will adjust itself to compensate for any slight warming that may occur. Like many who believe global warming poses a danger to humanity, Lindzen has often used his convictions to influence policy makers. In 1989, for example, Lindzen, Jerome Namias of Scripps Institution of Oceanography, and Reginald Newell of MIT wrote a letter to President George H. W. Bush to voice their support of a George C. Marshall Institute report concluding that the science offered by climatologists in support of carbon dioxide mitigation was too uncertain to justify action to reduce emissions.

Syukuro Manabe, Japanese physicist recruited by Joseph Smagorinsky in 1958 to work on the problem of accounting for the greenhouse effect in computer simulations of the atmosphere. One of Manabe's primary contributions to the project was to reformulate the general climate model to include the process of convection that occurs when greenhouse warming raises the temperature of the Earth's surface. In 1967, Manabe and his colleagues used this improved model to calculate that a doubling of atmospheric carbon dioxide would cause average global surface temperatures to increase by about 2.2 degrees Celsius (4 degrees Fahrenheit). In 1968, Manabe and co-researcher Richard Wetherald created the first three-dimensional computer model of the entire planet, which increased the predicted rise in average temperature to 3 degrees Celsius (5.4 degrees Fahrenheit) and calculated an 8 percent increase in evaporation of water in the atmosphere. These findings, published by Manabe and Wetherald in 1975, brought the idea that human activities could influence global climate to the attention of many scientists.

Patrick J. Michaels, research professor of environmental sciences at the University of Virginia and a fellow of the Cato Institute, which has published some of his writings. Skeptical of claims that global warming denotes imminent disaster for the planet, Michaels believes that the effects of climate change will be small and mostly beneficial. Although he has been a contributing author and reviewer of IPCC reports, he has questioned the validity of several of the international organization's conclusions. In 1998, while testifying before Congress, Michaels stated his belief that predictions of temperature change should be reduced from the then-current estimate of 2 degrees Celsius (3.5 degrees Fahrenheit) to just below 1 degree Celsius (1.8 degrees Fahrenheit).

Thomas Midgley, Jr., U.S. chemist who worked for General Motors Research Corporation during the early decades of the 20th century. Among his inventions was lead tetraethyl, an antiknock compound that increased the efficiency of engines while reducing wear. The use of this leaded gasoline, later found to be harmful to humans and to the environment, was banned with the advent of the catalytic converter. Midgley also invented chlorofluorocarbons (CFCs) for use as an artificial refrigerant. This compound of chlorine, fluorine, and carbon atoms was also found to be harmful to the environment as a destroyer of atmospheric ozone. As a result, CFCs were banned by the 1987 Montreal Protocol.

Milutin Milankovitch, Serbian mathematician who in the 1920s theorized that the coming and going of ice ages was determined by long-term cycles in the Earth's position and movement relative to the Sun. The three overlapping Milankovitch cycles, as they came to be known, repeat in periods of approximately 100,000 years, 41,000 years, and 20,000 years. Milankovitch's theory was confirmed in the 1970s and 1980s when clues in the fossil record revealed that the waxing and waning of past ice ages coincided with periodic wobbles in the Earth's rotation and variation in tilt of the Earth's axis.

John Mitchell, scientist at the British Meteorological Office's Hadley Center for Climate Prediction and Research who led a research group that succeeded in creating computer models accounting for the effects of aerosols in the atmosphere. In 1999, he helped Hadley Center colleague Simon F. B. Tett determine that global temperature rise in the early part of the 20th century could be attributed to a combination of increased solar radiation and greenhouse gases. However, since the 1970s, a period during which half of the observed warming has taken place, greenhouse gases have been the primary cause of global warming. In 2001, Mitchell replaced Benjamin Santer as the convening lead author of the IPCC assessment's chapter on the detection of the greenhouse fingerprint.

Mario J. Molina, scientist who in 1974, while working as a postdoctoral fellow at the University of California, Irvine, published an article with chemistry professor F. Sherwood Rowland claiming that CFCs destroy atmospheric ozone. Molina shared the 1995 Nobel Prize in chemistry with Rowland and Paul J. Crutzen for this work.

Ranga B. Myneni, professor of geography and researcher for the Climate and Vegetation Research Group at Boston University. Among Myneni's research has been an analysis of changes in the amount of solar radiation absorbed by plants north of the 45th parallel, which runs through Minneapolis, Boston,

Bordeaux, and Belgrade. He found that, during the decade between 1981 and 1991, the growing season had become about 12 days longer and that the amount of vegetation during the peak summer growing months of July and August had increased by about 10 percent. Myneni attributed these findings to a combination of global warming and the fertilization effects of increased amounts of carbon dioxide in the atmosphere.

Rejendra Pachauri, chairman of the IPCC. Pachauri was appointed in April 2002 after leading climate scientist Robert Watson was ousted from the position following the withdrawal of support by the U.S. government, whose stance had been heavily influenced by the oil company ExxonMobil. Pachauri, a respected science administrator, economist, and former railway engineer from India, became the first non–atmospheric chemist to hold the chairmanship of the IPCC.

Tim Palmer, climatologist at the European Center for Medium-Range Weather Forecasts in England. Palmer and co-researchers, after examining atmospheric circulation data for the second half of the 20th century, determined that the circulation patterns of the North Atlantic, North Pacific, and Arctic oscillations that caused warmer winters in the Northern Hemisphere had become more frequent, leading to unusually rapid warming in Alaska, Siberia, and western Canada. Palmer theorized that global warming had caused the increase in the frequency of these circulation patterns.

David Parker, research analyst at the British Meteorological Office who helped Chris Folland collect global ocean temperature data, the results of which were published in 1984 and which showed an upward trend in global temperature.

Camille Parmesan, a graduate student at the University of Texas at Austin who in 1992 began a rigorous four-year field study of the range of the Edith's checkerspot butterfly *(Euphydryas editha).* Her research showed that the butterfly's range had moved north and into higher altitudes, in accordance with what would be expected according to current global warming data.

Donald Pearlman, Washington, D.C.–based lawyer and lobbyist who represents U.S. oil and coal interests on the climate change issue. He has repeatedly attacked the validity of the IPCC's assessment reports.

Joyce Penner, professor of atmospheric, oceanic, and space sciences at the University of Michigan and director of the Laboratory for Atmospheric Science and Environmental Research. In collaboration with Karl E. Taylor, she used a general circulation model to research the combined effect of increased carbon dioxide and sulfate aerosol emissions on global temperatures. Their data indicated a general cooling of the stratosphere and a general warming of the troposphere, a pattern more likely to be caused by human-induced climate change than to occur naturally. Penner was appointed coordinating lead author of the IPCC Third Assessment Report's chapter on aerosols.

Veerabhadran Ramanathan, atmospheric chemist who published an article in October 1975 issue of *Science* in which he showed that chlorofluorocarbons (CFCs) trap the Sun's energy much more efficiently than water vapor, carbon dioxide, methane, and other greenhouse gases. He wrote that CFCs "may lead to an appreciable increase in the global surface temperature if the atmospheric concentrations of these compounds reach the values of the order of 2 parts per billion."

William K. Reilly, administrator of the Environmental Protection Agency (EPA) during George H. W. Bush's presidency. He urged strong steps to combat global warming, including the development of more efficient energy use.

Roger Revelle, director of the Scripps Institution of Oceanography in La Jolla, California, during the 1950s. In 1957 he collaborated with fellow Scripps oceanographer Hans Suess on a study that reached the conclusions that atmospheric carbon dioxide was rising and that humans were "carrying out a large scale geophysical experiment of a kind that could not have happened in the past nor be reproduced in the future."

Rebecca J. Ross, research meteorologist at the National Oceanic and Atmospheric Administration (NOAA) who in 1996 published a study showing that the amount of water vapor in the atmosphere over the United States had increased by 3 to 7 percent per decade in recent decades, a change that may have resulted from global warming. In 1998, Ross joined fellow NOAA researcher Dian J. Gaffen in a study that showed an increased frequency of extremely hot, humid days in the United States between 1949 and 1995.

Carl-Gustaf Rossby, Swedish meteorologist who was a member of Norway's Bergen School during the early decades of the 20th century. He conducted pioneering research into the behavior of cold and warm air masses and discovered what became known as Rossby waves, a description of airflow behavior in the jet stream. Rossby moved to the United States in 1929 and became a citizen in 1939. He was instrumental in developing well-respected meteorology departments at MIT and the University of Chicago and also helped raise funds for the founding of the Meteorology Project at the Institute for Advanced Study in Princeton, New Jersey. In 1950, seven years before his death, he returned to Sweden to help found the Institute of Meteorology at the University of Stockholm.

F. Sherwood Rowland, the Donald Bren Research Professor of Chemistry and Earth System Science at the University of California, Irvine. In June 1974, Rowland and Mario J. Molina published a groundbreaking article in *Nature* contending that chlorofluorocarbons (CFCs) destroy atmospheric ozone. Rowland's continuing research into ozone resulted in the 1995 Nobel Prize in chemistry, which he shared with Molina and Paul J. Crutzen.

Benjamin D. Santer, research scientist at the Lawrence Livermore National Laboratory. As a doctoral student in climate science at the University of East Anglia, England, in the mid-1980s, he conducted early research into the use of formal statistical methods to test the accuracy of general circulation models. In 1987, Santer began working at the Max Planck Institute for Meteorology in Hamburg, Germany, where his research led him to conclude that geographical patterns of surface temperature and the pattern of temperature variation from the bottom to the top of the atmosphere provided the best clues to detecting the signal of anthropogenic climate change amid the confusion of natural variability. In 1992, he joined Lawrence Gates's climate research group at the Lawrence Livermore National Laboratory, east of San Francisco, and continued his attempts to detect a definitive greenhouse fingerprint. Santer was named the convening lead author of the 1995 IPCC report's chapter dealing with the greenhouse fingerprint. After the report's publication, he was accused by fossil fuel industry representatives and global warming skeptics of rewriting the chapter without proper authority. The attack backfired when other scientists flocked to Santer's defense and vindicated his methods.

Michael E. Schlesinger, research scientist who, in the late 1980s, helped Tim Barnett determine that tropospheric temperature, atmospheric pressure at sea level, and sea-surface temperature were more reliable as greenhouse gas indicators than most other variables.

Stephen S. Schneider, climatologist who, during the 1970s, believed that human-made pollutants in the atmosphere were preventing sunlight from reaching the Earth, resulting in a long-term global cooling trend. He later changed his mind and supported the view that greenhouse emissions were causing the Earth to warm. Although global warming skeptics have pointed to this reversal as evidence that the science of climatology is too uncertain to be used as the basis for policy decisions, Schneider has maintained that the ability to change one's mind in the face of new evidence is the "ultimate test" of respectability for scientists.

Frederick Seitz, president of the National Academy of Sciences during the 1960s, chairman of the George C. Marshall Institute, and president emeritus of Rockefeller University. Though he is not a climate scientist, Seitz, a physicist, is a global warming skeptic who has taken part in attacks on the credibility of the Intergovernmental Panel on Climate Change. In an article published in the *Wall Street Journal* in 1996, he accused Benjamin Santer of corrupting the peer-review process by changing sections of the IPCC assessment without proper authority.

S. Fred Singer, professor of environmental sciences at the University of Virginia and founder of the Science and Environmental Policy Project. After earning his doctorate in physics from Princeton University, Singer conducted pioneering research into the development of rocket and satellite technology during the 1950s and 1960s. In 1962, he was appointed the first director of the national Weather Satellite Center. In response to James Hansen's 1988 congressional testimony concerning the dangers of global warming, Singer founded a think tank called the Science and Environmental Policy Project (SEPP) in 1990 to espouse the opposing point of view. He has since worked to prove that the Earth has not experienced and will not undergo significant warming and that any small rise in global temperature is more likely to benefit humans than harm them.

Hans Suess, oceanographer at the Scripps Institution of Oceanography in La Jolla, California, during the 1950s. He collaborated with Scripps director Roger Revelle on a research project, the results of which were published in 1957, to study the exchange of carbon dioxide between the atmosphere and the oceans and to determine whether the amount of atmospheric carbon dioxide had increased during the previous decades. The article concluded that humans were "carrying out a large scale geophysical experiment of a kind that could not have happened in the past nor be reproduced in the future."

Karl E. Taylor, scientist at Lawrence Livermore National Laboratory in California who helped Joyce Penner use computer models of the atmosphere to investigate the combined effect of global temperature of increasing carbon dioxide and sulfate aerosol emissions. Taylor also contributed to the IPCC's Second Assessment Report in 1995.

James Tyndall, Irish mountain climber and glacier researcher who is credited with coining the term "greenhouse gases." He published a paper in 1861 in which he measured and demonstrated the high heat radiation absorption powers of atmospheric water vapor, carbon dioxide, and ozone. He concluded that lower levels of atmospheric carbon dioxide would result in another ice age, but he failed to consider the potential warming effects of increased levels of greenhouse gases.

John Von Neumann, Hungarian-born mathematician who moved to Princeton, New Jersey, in 1930 to work at the recently opened Institute for Advanced Study. He is credited with designing one of the first electronic computers, the use of which he applied to weather forecasting after founding the Meteorology Project with the help of Carl-Gustaf Rossby.

In 1950 Von Neumann, Jule Charney, and other researchers launched the first computerized simulation of the weather.

Robert Watson, chairman of the Intergovernmental Panel on Climate Change (IPCC) from 1999 to 2002. During his tenure, Watson was an outspoken advocate of action to combat global warming. As a result, the United States formed an alliance with many developing countries to oust Watson and award the IPCC chairmanship to India's Rejendra Pachauri, who was known to be less resistant toward fossil fuel industry interests.

Richard Wetherald, researcher at the U.S. federal government's Geophysical Fluid Dynamics Laboratory (GFDL) in Washington, D.C., in the 1960s who helped Syukuro Manabe develop the first three-dimensional computer model of Earth. The model was used to predict a 3-degree-Celsius (5.4-degree-Fahrenheit) rise in average global temperature and an 8 percent increase in evaporation of water in the atmosphere, findings that were published in 1975.

Christine Todd Whitman, head of the U.S. Environmental Protection Agency (EPA) during the George W. Bush administration. As governor of New Jersey from 1993 to 2000, she earned a mixed reputation among environmentalists, who applauded her support of a measure earmarking $1 billion to preserve land from development but criticized attempts to reduce fines for polluters. She called global warming "one of the greatest environmental challenges we face, if not the greatest" but condemned the release of an EPA report concluding that greenhouse gas emissions produced by human activities are the primary cause of climate change.

Thomas M. L. Wigley, researcher at the University of East Anglia in Norwich, England, who worked with colleague Phil Jones and analysts at the British Meteorological Office during the early 1980s to collect and combine land temperature records from around the world. He was appointed one of the lead authors of the 1995 IPCC scientific report. Wigley also became a mentor to Benjamin Santer while the latter worked toward his doctorate in East Anglia. The two collaborated on attempts to detect definitive signals of anthropogenic climate change.

Timothy Wirth, U.S. senator from Colorado who scheduled the June 23, 1988, hearing at which James Hansen brought global warming into the forefront of public and political consciousness. During the Clinton administration he served as chief U.S. delegate to the IPCC during early attempts to draft an international agreement to reduce greenhouse gas emissions.

CHAPTER 5

GLOSSARY

Global warming studies combines research from a variety of scientific fields—including meteorology, climatology, chemistry, and physics—each with its own specialized terminology. This chapter presents many of the terms that students are likely to encounter while conducting general research into climate change, global warming, and related environmental issues. Several web sites also offer extensive online glossaries (see Chapter 6, "How to Research Global Warming").

acid rain Rain made acidic by combination with atmospheric sulfur dioxide and nitrogen oxide. Acid precipitation also occurs in the form of fog and snow. In the United States, approximately two-thirds of all sulfur dioxide and one-quarter of all nitrogen oxide comes from the burning of fossil fuels used to generate electricity. Acid rain, upon reaching the Earth, affects a variety of animals (including fish), trees, and other living things that rely on the water. Acidic precipitation is also capable of reacting with metals to form corrosive salts.

aerosols Any small airborne particles. Sulfur dioxide particles, known as sulfate aerosols, are created naturally by volcanic eruptions and anthropogenically by the burning of fossil fuels. They reflect sunlight back into space, thereby cooling the Earth and slightly masking the effects of global warming. Sulfate aerosols also contribute to the problem of acid rain.

Agenda 21 The action plan signed by more than 160 nations at the 1992 United Nations Conference on Environment and Development (UNCED) in Rio de Janeiro, Brazil, on a wide range of environmental, development, social, and economic issues to be dealt with during the 21st century.

albedo The reflection of incoming solar light into space by clouds, aerosols, the atmosphere, ice, water, and land surface, also known as reflectivity. Highly reflective surfaces such as clouds, ice, and snow are said to have a high albedo, while plants, which absorb light for photosynthesis, have a low albedo.

Glossary

Antarctic The area surrounding the South Pole, usually considered to be the region south of 66° 32′ South latitude.

anthropogenic effects Changes in the environment caused by human activities. Global warming is considered by many scientists to be an anthropogenic effect caused by the burning of fossil fuels by humans.

aquifer An underground geological formation that acts as a reservoir for water, which may be tapped by humans for agricultural, industrial, recreational, and domestic use. In some areas, the use of water from aquifers exceeds the rate of replenishment by rainwater and snowmelt, endangering the long-term survival of surrounding communities. Some scientists worry that changes in precipitation and drought patterns caused by global warming may affect the viability of important aquifers.

Arctic The area surrounding the North Pole, usually considered to be the region north of 66° 32′ North latitude.

atmosphere The blanket of gases, approximately 560 kilometers (348 miles) thick, that surrounds the Earth. It is primarily composed of nitrogen (79 percent) and oxygen (20.9 percent), with trace amounts of water, ozone, carbon dioxide, and other greenhouse gases. The four distinct layers of the atmosphere are, from the ground up, the troposphere, the stratosphere, the mesosphere, and the thermosphere.

atmosphere-ocean general circulation model (AOGCM) *See* **general circulation model (GCM).**

average In climatic terms, the average occurrence of weather-related phenomena such as precipitation and temperature over a period of 30 years.

Berlin Mandate A decision reached at the 1995 United Nations Berlin Climate Summit in which participating nations agreed that developed nations should strengthen their commitment to the 1992 United Nations Framework Convention on Climate Change (UNFCCC) by committing to quantified reductions of greenhouse gas emissions within specified time frames. The Berlin Mandate was the precursor to the 1997 Kyoto talks.

biocentrism *See* **deep ecology.**

biodiversity (biological diversity) A measure of the variety and genetic diversity of different biological species found in a particular area. Many scientists believe that rapid changes in ecosystems brought about by global warming may kill off vulnerable species, thus reducing biodiversity in affected areas.

biological pump A process that begins when atmospheric carbon dioxide is used for photosynthesis by phytoplankton. While most of the carbon dioxide is released back into atmosphere, a portion sinks to the bottom of the ocean in the remains of decaying organic matter, thus removing carbon from the carbon cycle for periods that may reach millions of years. This process is important because it regulates the amount of carbon dioxide in both the deep oceans and in the atmosphere. Many researchers believe that, without such a biological pump, the concentration of carbon dioxide in the atmosphere would more than double.

biomass The amount of plant life in a particular region. The burning of biomass, which releases carbon dioxide into the atmosphere, constitutes the primary energy source for approximately one-half of the world's population.

biome The community of plants and animals living in a common, specified natural region such as a rain forest, desert, or tundra.

biosphere Collectively, the regions of the Earth that are capable of supporting living organisms. This includes areas on land, in the oceans, and in the atmosphere.

boreal forests Wooded areas in far northern latitudes consisting of evergreen trees. Many scientists worry that climate change will vastly reduce the range of such forests because the trees will be unable to migrate to cooler regions.

business-as-usual scenario Term used to denote a future in which no major action is taken by the global community to reduce the emission of greenhouse gases, such as shifting energy use patterns from fossil fuels to sustainable sources.

butterfly effect The idea, often cited in chaos theory, that a single butterfly flapping its wings in, for example, Japan can conceivably set in motion a chain of unpredictable climatic flips that could eventually trigger a hurricane in, for example, North America. This theory is used to illustrate the impossibility of predicting weather and climate over the long term.

carbon cycle The process by which carbon, in the form of carbon dioxide, is exchanged among natural carbon reservoirs, including the atmosphere, the oceans, and vegetation.

carbon dioxide (CO_2) A molecule consisting of one atom of carbon and two atoms of oxygen. The greenhouse gas of greatest concern in the study of global warming, anthropogenic carbon dioxide, is released into the atmosphere mainly through the burning of fossil fuels and through deforestation.

carbon sequestration The absorption and storage of carbon released from another part of the carbon cycle. During the process of photosynthesis, for example, vegetation absorbs carbon dioxide from the atmosphere, then releases the oxygen and stores the carbon. Reservoirs in which the carbon is stored—such as forests, oceans, and unburned fossil fuels—are known as carbon sinks. Storage of carbon in such sinks can partially offset the effects of anthropogenic carbon dioxide emissions.

carbon sink *See* **carbon sequestration.**

carbon tax A fee placed on industrial emissions of carbon dioxide into the atmosphere.

Glossary

chlorofluorocarbons (CFCs) A set of synthetic compounds classified as greenhouse gases that have been used extensively for refrigeration, for aerosol sprays, and as industrial solvents. Because these compounds destroy ozone in the stratosphere and have an extremely long life in the atmosphere, their production and use has been drastically scaled down as a result of the 1987 Montreal Protocol.

climate The long-term weather conditions in a particular region, including such elements as precipitation, temperature, humidity, and wind velocity.

climatology The study of long-term weather conditions.

clouds Atmospheric phenomena consisting of vast quantities of tiny water droplets or ice crystals that form when rising and cooling air can no longer hold all the moisture within it as water vapor. Growth of the droplets within a cloud under the proper conditions may result in precipitation. Clouds play a dual role in the natural greenhouse effect: While they reflect some of the Sun's radiation back into space before it reaches the Earth's surface (cooling effect), they also absorb and emit heat that radiates from the Earth's surface (warming effect). The net effect of these two processes results, on average, in a slight cooling of the Earth's surface.

computer model *See* **general circulation model (GCM).**

convection The upward movement of air that has been heated by its contact with the Earth's surface or the ocean.

cryosphere The portion of the Earth's surface covered by ice and snow, including continental ice sheets, glaciers, snow cover, and sea, lake, and river ice.

cyclic pattern The theory that the Earth's climate varies according to long-term repetitions such as the sunspot cycle.

deep ecology An environmental philosophy in which all life on Earth has equal standing with humans and has the right to function normally, without human interference, in its ecosystem. It is sometimes referred to as biocentrism.

deforestation The large-scale logging of trees. Deforestation is one of the primary anthropogenic causes of global warming, not only because burning and decomposing vegetation releases sequestered carbon dioxide, but also because living trees act as a carbon sink by removing carbon dioxide from the atmosphere during the process of photosynthesis.

desertification The spread of arid desert areas into adjoining semi-arid regions. This process may be caused by temporary or long-term drought conditions, or by human activities such as overgrazing by livestock, the removal of natural vegetation, or the overuse of the soil through poorly managed agricultural operations.

ecosystem A distinct system of interdependent plants and animals, along with their physical environment.

El Niño Spanish for "Christ child," a weather phenomenon that occurs in the southeastern Pacific Ocean at irregular intervals every two to seven years during the Christmas season. It consists of seasonal changes in the direction of Pacific winds and abnormal warming of ocean water. A strong El Niño can affect global weather patterns.

enhanced greenhouse effect Additional global warming caused by greenhouse gases emitted into the atmosphere through human activity, such as the burning of fossil fuels and deforestation. Increased concentrations of these gases trap more infrared radiation, further warming the Earth's climate.

environmental refugees People forced to leave their homes because of adverse environmental conditions, such as droughts, floods, and sea level rise.

epoch A long-term measure of geological or climatic time.

evaporation The process whereby liquid water changes into invisible, gaseous water vapor and is taken up into the atmosphere.

feedback A process in which two interacting events influence each other. Feedback can be positive, in which the events enhance each other, or negative, in which the interacting events dampen the effects of both or cancel each other out.

feedback mechanism A mechanism that connects one system to another. A mechanism that enhances both systems is a positive feedback, while one that moderates or cancels out both systems is a negative feedback.

fertilization effect The process in which vegetation grows more rapidly and abundantly in an atmosphere that contains a greater concentration of carbon dioxide.

flip A sudden jump or change in climate theorized to occur as part of the butterfly effect and thought to make climate impossible to predict over the long term.

fossil fuel Any energy source created from the decayed remains of ancient plant and animal life, the burning of which releases carbon dioxide into the atmosphere. Among the fossil fuels are coal, natural gas, and oil.

general circulation model (GCM) Sometimes called atmosphere-ocean general circulation models (AOGCMs), computer models that simulate the interaction of the Earth's atmosphere and oceans and the manner in which they change over time. Such models are used to predict how the climate may be affected by the increasing amounts of greenhouse gases in the atmosphere. Critics claim that the interaction between the atmosphere and the oceans is much too complex for GCMs to accurately simulate and predict.

geoengineering Artificial modification of the environment to counteract the effects of global warming. Proposals have included injecting sulfur particles into the stratosphere, installing giant mirrors in space, and floating huge pieces of white plastic in the oceans, all intended to reduce incoming solar radiation.

geothermal energy Renewable energy obtained by transferring heat from deep down in the Earth's crust to the surface.

global warming The rise in the Earth's global temperature caused by an anthropogenic increase in the amount of greenhouse gases, particularly carbon dioxide, in the atmosphere.

greenhouse effect A naturally occurring process in which concentrations of greenhouse gases trap heat within the Earth's atmosphere, keeping the planet warm enough to sustain life. Greenhouse gases allow solar radiation to reach the Earth's surface but retain a portion of this heat in amounts directly related to increased concentrations of these gases in the atmosphere.

greenhouse gases The gases that allow solar radiation to reach the Earth's surface but efficiently absorb radiation that has been reflected back into space from the planet's surface, thereby contributing to the greenhouse effect. The primary greenhouse gases are water vapor, carbon dioxide, methane, nitrous oxide, CFCs, and ozone.

hurricane A powerful tropical or subtropical storm that originates during the summer or autumn in the region between five and 20 degrees north and south of the equator. During these extreme weather events, winds can reach speeds of up to 320 kilometers per hour (200 miles per hour), causing massive property damage and leading to loss of life. Many researchers worry that global warming may lead to changes in hurricane patterns, making them stronger or more frequent, or increasing the length of hurricane season.

hydropower Also called hydroelectric power, a renewable energy source in which flowing water is used to generate electricity.

hydrosphere The total portion of the Earth consisting of water, including oceans, seas, ice caps, glaciers, lakes, rivers, and underground reservoirs.

ice core An ice sample consisting of a tube drilled out from a glacier or an ice sheet. Scientists use ice cores to study gas bubbles, chemicals, and dust flecks trapped in the ice up to 250,000 years ago, which provide historical clues about the composition of the atmosphere, volcanic eruptions, precipitation, sudden shifts in temperature, and other climatic phenomena.

Industrial Revolution The period during the 18th and 19th centuries characterized by the rapid growth of industry, and the attendant increase in the burning of coal for energy, in Europe and North America. Researchers date the relatively recent rise in the concentration of carbon dioxide in the atmosphere to the beginning of the Industrial Revolution.

interglacial period A period of warm climatic conditions between successive ice ages that occurs roughly every 100,000 years. The Earth is currently in an interglacial period.

Intergovernmental Panel on Climate Change (IPCC) A multinational group of about 2,500 scientists formed in 1988 to assess the state of global warming and report its findings to governments and policy makers. Sponsored by the United Nations, the IPCC is widely considered to be world's foremost authority on the subject. The organization released major assessment reports about the state of the global warming problem in 1990, 1995, and 2001.

jet stream A zone of extremely strong winds that flows in the upper troposphere.

Kyoto Protocol A document drafted in 1997 in Kyoto, Japan, in which industrial and former Eastern bloc nations agreed to reduce, during the period from 2008 to 2012, greenhouse gas emissions by 5.2 percent below 1990 levels.

La Niña A weather phenomenon during which ocean temperatures in the southeastern Pacific Ocean grow cooler. As with El Niño, a strong La Niña can influence global weather patterns.

Little Ice Age A period lasting from about A.D. 1300 to 1900 during which average temperatures in the North Atlantic region dropped by about 1.5 to 2 degrees Celsius (2.7 to 3.6 degrees Fahrenheit), leading to extensive glaciation and shorter agricultural growing seasons in the Northern Hemisphere.

Medieval Warm Period Also known as the Little Climatic Optimum, a period lasting from A.D. 900 to 1300 during which temperatures in the Northern Hemisphere were slightly warmer and rainfall was abundant, supporting a rapid expansion of agriculture in Europe.

mesosphere The portion of the atmosphere that starts just above the stratosphere, extending from 50 kilometers (31 miles) to 85 kilometers (53 miles) above the Earth's surface.

microclimate The long-term weather conditions in a very small area, such as a single field or valley.

Milankovitch cycle Based on a theory developed by Serbian mathematician Milutan Milankovitch, three overlapping cycles of warm and cold temperatures on Earth lasting 100,000, 41,000, and 20,000 years respectively. According to Milankovitch, these fluctuations are linked to regular variations in the Earth's orbit around the Sun that cause fluctuations in the distribution of incoming solar radiation.

monsoon A period of heavy rainfall that occurs in subtropical regions between April and December of each year.

ozone An unstable greenhouse gas that occurs at two distinct layers in the atmosphere: in the low-altitude troposphere, where it exists as a form of air pollution produced by automobile emissions, and in the high-altitude stratosphere, where it is created naturally by sunlight. The stratospheric ozone layer shields the Earth from dangerous ultraviolet radiation from the Sun.

ozone hole A region of the atmosphere over Antarctica where about one-half of the stratospheric ozone disappears every spring (August to October in the Southern Hemisphere), only to return during the summer months (December and January). The hole, which allows increased concentrations of cancer-causing ultraviolet radiation to reach the Earth's surface, was first detected in 1985 and was determined to have been created by human-made chlorofluorocarbons (CFCs) in the atmosphere.

paleoclimatology The study of long-term weather conditions of the past by such means as ice-core measurements or the examination of tree rings.

photosynthesis The process by which plants use energy from the Sun to extract carbon dioxide from the atmosphere and convert it into plant tissue. During the conversion, oxygen is released into the atmosphere, while carbon remains stored in the plant tissue.

photovoltaic (PV) cell An energy cell that converts solar radiation into electricity.

phytoplankton Plankton that consists of minute forms of plant life.

plankton Minute, generally microscopic, plant and animal organisms that float in the ocean.

precipitation Water that falls to the Earth's surface from clouds, which may, depending upon temperature and other variables, occur in the form of rain, hail, snow, sleet, or dew.

radiation Energy in the form of invisible electromagnetic waves that travel at the speed of light.

radiation budget The balance of radiation that enters and leaves the Earth's atmosphere. In a balanced system, in which average global temperatures will remain relatively stable, the quantity of solar radiation entering the atmosphere from space is roughly equivalent to the thermal radiation leaving the Earth's surface and atmosphere.

radiative forcing Changes to the Earth's balance of incoming and outgoing radiation (radiation budget), caused, for example, by an increase in the concentration of atmospheric carbon dioxide or by variations in the release of energy from the Sun. Such changes may affect average global temperatures.

rainfall The total depth of measured precipitation, including such forms as snow and hail, in any given location.

reflectivity *See* **albedo**.

reforestation The replanting of trees in areas where forests had previously been cleared. Massive reforestation may help mitigate the effects of global warming by creating new carbon sinks, which will remove additional carbon dioxide from the atmosphere during the process of photosynthesis.

renewable energy Energy sources that are not depleted by use. Among these are hydropower, solar power, and wind power.

runaway greenhouse effect An exaggerated greenhouse effect in which an atmosphere that contains a high density of greenhouse gases results in additional warming, which in turn contributes to other factors that enhance global warming. The planet Venus, where temperatures reach 470 degrees Celsius (896 degrees Fahrenheit), provides a good example of the results of a runaway greenhouse effect.

salinity A measure of the amount of salt in ocean water.

shelf ice Ice in the sea that is attached to a land glacier.

smog Pollution caused by emissions from automobiles and industry. Tropospheric ozone is an element of smog.

solar power Renewable energy derived from solar radiation.

solar radiation Energy emitted by the Sun, consisting of about 47 percent infrared radiation, 46 percent invisible radiation, and 7 percent ultraviolet radiation.

stratosphere The region of the atmosphere just above the troposphere that extends from about 10 kilometers (6 miles) to 50 kilometers (31 miles) above the Earth's surface. The ozone layer, which protects the planet from ultraviolet radiation, is contained within in the stratosphere.

sunspot A dark region in the Sun's photosphere of higher temperature and intense magnetism. These regions, which usually occur in groups, typically measure 2,000 to 3,000 kilometers (1,100 to 1,700 miles) across and last only two or three weeks before disappearing. Theories that sunspot activity may influence temperatures on Earth have been presented by global warming skeptics to explain weather changes that have otherwise been attributed to greenhouse gases in the atmosphere.

terrestrial radiation Energy emitted by the Earth, consisting mostly of infrared radiation.

thermal expansion The increase in water volume caused by increasing water temperature. Scientists believe that this, and not melting glaciers, will be responsible for most of the sea level rise that may accompany global warming.

thermosphere Also known as the upper atmosphere, the layer of the atmosphere that starts just above the mesosphere at 85 kilometers (53 miles) above the Earth's surface and extends to 600 kilometers (372 miles) high. Because of the Sun's energy, temperatures rise as altitude increases.

tornado An extreme weather phenomenon consisting of a small area, often of less than 500 meters (1600 feet), of strong upward air currents. Tornadoes are capable of causing severe property damage and loss of life.

troposphere The region of the lower atmosphere that extends from the Earth's surface up to a height of about 10 kilometers (6 miles). It is in this layer, where temperature falls as altitude increases, that most weather occurs.

United Nations Conference on Environment and Development (UNCED) Also known as the Earth Summit, a meeting held in Rio de Janeiro, Brazil, in June 1992 that resulted in Agenda 21 and the United Nations Framework Convention on Climate Change (UNFCCC), which was signed by 160 participating countries.

United Nations Framework Convention on Climate Change (UNFCCC) Introduced at the Earth Summit in 1992 and entered into force in March 1994, the UNFCCC sought to stabilize atmospheric concentrations of greenhouse gases without compromising economic development. The treaty established a voluntary goal of returning greenhouse emissions to 1990 levels by the year 2000. Many signatories considered this goal inadequate and sought legally binding commitments, an effort that culminated in the 1997 Kyoto Protocol.

United Nations World Summit on Sustainable Development Also known as the Johannesburg Summit (JOWSCO), a meeting held in Johannesburg, South Africa, in 2002 at which more than 100 heads of state gathered to draft a plan to implement the environmental goals outlined at the 1992 Earth Summit. The resulting Johannesburg Implementation Plan set an ambitious, nonbinding agenda to deal with issues ranging from clean water and toxic chemicals to poverty and global warming.

urban heat island Phenomenon in which the temperatures of densely populated areas are warmer than surrounding rural areas because of the heat-retention effect of tightly clustered buildings.

vertical overturning The horizontal circulation of ocean water, by which water rises to the surface and then sinks to the depths in repeating cycles.

wind power Renewable energy collected from turbines or windmills that turn under windy conditions.

Younger Dryas A climatic event that occurred about 12,000 years ago and lasted for about 1,500 years, during which cold temperatures and reglaciation suddenly interrupted the warming of the Earth following the last ice age. The Younger Dryas (named after an arctic flower that spread during the period) ended just as quickly as it began, with temperatures rising in 5.5-degree-Celsius (10-degree-Fahrenheit) jumps in less than a decade. This event provides evidence that the Earth's climate, rather than changing gradually, is capable of massive fluctuations during short time periods.

zooplankton Plankton that consists of minute forms of animal life.

PART II

GUIDE TO FURTHER RESEARCH

CHAPTER 6

How to Research Global Warming

Before beginning any research project, it is important to have a topic or theme, a list of key terms and searchable words, and a general idea of how to use the variety of research tools. This chapter suggests a number of web sites, indexes, and catalogs that will help organize and define the research process and will provide a wide range of information, from a variety of points of view, pertaining to global warming and climate change.

THE INTERNET

The Internet and the World Wide Web offer virtually unlimited amounts of information on global warming. Keep in mind, however, that web sites reflect the agenda and biases of their creators. This is particularly important to keep in mind when visiting sites created by environmentalist groups and public policy research organizations, both of which are capable of manipulating or exaggerating statistics to support their point of view. Many public policy research organizations claim to present unbiased information when, in fact, they adhere to a particular political agenda. The authors of these web sites often declare that their information is based on rigorous, objective scientific research, then go on to blame the opposition of "politicizing" the debate for their own ends. Realize, however, that there are political elements on all sides of the global warming debate; any information should therefore be weighed against opposing points of view supplied by other web sites and non-Internet resources. For this reason, it is also important, when using statistical information in a report, to cite the source in case questions of validity arise.

Like any research tool, the Internet is most useful if time is taken to learn the various tricks that will allow the researcher to efficiently tap into all the resources and information it has to offer. It is important to remember, however, that although the World Wide Web is a remarkable resource, it is best used as a supplement to, rather than as a replacement for, research materials that are available only by visiting a library.

Unlike books, newspapers, periodicals, and video documentaries, the Internet is a nonlinear research medium, meaning that the information is not offered in a straight line. The user, rather than the author or producer, decides where to search for information and in what order the information will be received and reviewed. A single web site can provide links to a number of other sites, each of which may, in turn, lead to additional sites. One site that does not appear relevant to the topic at hand may provide access to one or more sites that deal precisely with the subject matter being researched. Therefore, diligence is important. Navigating in this manner can provide large amounts of information, but following a seemingly endless array of links can also cause the researcher to lose track of useful pages. Most web hosting services aid in research organization by offering the ability to store "favorite" or "bookmarked" sites that are likely to be revisited in the future. The exact method of bookmarking a web site may vary among Internet providers, but generally there is an onscreen icon that will save the web page being viewed to a "favorites" folder. Another useful feature is the "history" menu, which is a list of all sites visited during a research session or in the recent past. This acts as a sort of "breadcrumb trail" that allows surfers to retrace their steps or revisit any site or link that has been hit along the way.

"Surfing the web" by following links can, as has already been mentioned, lead to large amounts of information, but this inherently haphazard method can also cause researchers to miss important web sites altogether. Web indexes and search engines can fill in these information gaps.

INDEXES

A web index, or guide, is a site that offers a structured, hierarchical listing or grouping of key terms or subject areas relating to the topic requested. This allows researchers to focus on a particular aspect of a subject and find relevant links and web sites to explore.

Web indexes possess several advantages over random or blind Internet surfing. First, they offer a structured hierarchy of topics and terms that simplifies the process of honing in on a specific topic, related subtopic, or link. Second, sites are screened and evaluated for their usefulness and quality by those who compile the index, giving the researcher a better chance of finding more substantial and accurate information. This feature also has its down side, however, which is that the index user is at the mercy of the indexer's judgment about which sites are worth exploring. As with all other research tools, web indexes should therefore be used in conjunction with other tools and methods.

One of the most popular and easiest-to-use web indexes is Yahoo! (http://www.yahoo.com). Researchers can use the homepage's top-level list of topics and follow them to more specific areas, or they can type one or more key words into the search box and receive a list of matching categories and sites. To explore climate change via Yahoo!, the researcher could click on the "Society and Culture" link, then "Environment and Nature." From there, the user can choose such subtopics as "Climate Change," "Energy," "Environmental Economics," "Global Change," "Global Warming," "Meteorology," "Oil and Gas Issues," "Pollution," or "Sustainable Development," depending on the focus of the research. Global warming can also be explored through the "Science" link, followed by "Ecology" or "Meteorology," and then a number of subtopics. These topics are just suggestions to get started, of course. Since global warming encompasses a number of scientific, social, economic, and political disciplines, it is useful to keep an open mind about where information can be found. Do not limit research to the obvious, and supplement browsing with a direct (random) search to ensure the most comprehensive results.

About.com (http://www.about.com) is similar to Yahoo! but gives greater emphasis to guides prepared by "experts" in various topics. Information on global warming can be found by browsing "News and Issues," then "Environmental Issues" (offering such links as "Air Pollution," "Renewable Energy," and "Climate Change"), "Global Issues" (then "Environment"), "Conservative Politics: U.S." (then "Environment"), or "Liberal Politics: U.S." (then "Environment").

Another example of a useful web index is AskJeeves (http://www.ask.com or http://askjeeves.com). This site attempts to answer plain-English questions, such as "What is global warming?" Sometimes it directly answers the question, and other times it provides a number of possibly useful links it obtains by scanning a series of search engines.

SEARCH ENGINES

When beginning an online research project, several organizations' web sites may come to mind, but there are many more out there that may be lesser known but are no less valuable. This is where search engines come into the picture. A search engine scans web documents for key words or terms provided by the researcher and comes up with a list of relevant web sites to explore. Instead of organizing topically in a top-down fashion, search engines work their way from the bottom up, meaning that they search the web for key terms and compile their findings into an index. Next, the search engine takes that index and matches the search words to those links that have been flagged as key term matches. Finally, the engine compiles a list based on the sites within the index that match the entered searchable words.

There are hundreds of search engines. Among the most user-friendly and popular are

Alta Vista (http://www.altavista.com)
Excite (http://www.excite.com)
Go (http://www.go.com)
Google (http://www.google.com)
Hotbot (http://hotbot.lycos.com)
Lycos (http://www.lycos.com)
Northern Light (http://www.northernlight.com)
WebCrawler (http://www.webcrawler.com)

Search engines are easy to use by employing the same kinds of key words that work with web indexes and library catalogs. A variety of web search tutorials are available online.

There are some basic rules for using search engines. When looking for something, use the most specific term or phrase. For example, when researching atmospheric carbon dioxide, using the phrase "carbon dioxide emissions" will result in more useful information than simply entering "air pollution." Note that phrases should be placed in quotation marks in the search field if you want them to be matched as phrases rather than as individual words.

When searching for a more general topic, use several descriptive words (nouns are more reliable than verbs), such as "global climate change." Most search engines will automatically put pages that match all three terms first on the results list.

Use "wildcards" when a search term may have more than one ending. For example, typing in "atmospher*" will match both "atmosphere" and "atmospheric."

Most search engines support Boolean (and, or, not) operators that can be used to broaden or limit a search topics. Use AND to narrow a search: "oil AND gas" will match only pages that have both terms. Use OR to broaden a search: "oil OR gas" will match any page that has either term. Use NOT to exclude unwanted results: "fossil fuel NOT gas" finds articles about fossil fuel except those relating to gas.

Each search engine has its own method of finding and indexing results and will therefore come up with a unique list. It is therefore a good idea to use several different search engines, particularly for a general query. Several "metasearch" programs automate the process of submitting search terms to multiple search engines. These include Metacrawler (http://www.metacrawler.com) and Search.com (http://www.search.com).

FINDING ORGANIZATIONS AND PEOPLE

Web sites of organizations can often be found by entering the name into a search engine. Generally, the best approach is to put the name of the organization into quotation marks, such as "Climate Solutions." If this does not yield satisfactory results, another approach is to take a guess at the organization's likely web address. Climate Solutions' site, for example, can be found by typing in www.climatesolutions.org. The National Aeronautics and Space Administration is commonly known by the acronym NASA, so it is no surprise that this government agency's web site is www.nasa.gov. (Keep in mind that noncommercial organization sites normally use the *.org* suffix, government agencies use *.gov*, educational institutions use *.edu*, and businesses use *.com*).

There are several ways to find people on the Internet. Entering the person's name (in quotes) in a search engine may lead to that person's homepage on the Internet. Another way is to contact the person's employer, such as a university for an academic or a corporation for a technical professional. Most such organizations have web pages that include a searchable faculty or employee directory. Finally, one of the people-finder services, such as Yahoo! People Search (http://people.yahoo.com) or BigFoot (http://www.bigfoot.com), may yield an e-mail address, regular address, and/or phone number.

SPECIFIC RESOURCES

A variety of government, educational, and private web sites offer background material, news, analysis, and other materials on topics, groups, and people associated with global warming research. The following are some of the more useful major sites, broken down by category.

GENERAL WEATHER SITES

- Local weather forecasts and general meteorological information can be found on most newspaper web sites. For national and international data, useful resources include CNN Weather's home page (http://www.cnn.com/weather), USA Today Weather (http://www.usatoday.com/weather/wfront.htm), and The Weather Channel (http://www.weather.com).
- Live Weather Images (http://weatherimages.org) pulls together the most useful and frequently accessed weather data on the Internet. The site provides access to current weather images and forecasts, a live message board,

a newsletter, and an interactive weather page where web surfers can calculate sunrise/sunset times, heat indexes, wind chill factors, and more.

- NASA's Geostationary Operational Environmental Satellite (GOES) Project Science (http://rsd.gsfc.nasa.gov/goes) monitors the Western Hemisphere for unpredictable weather, especially hurricanes. The site provides up-to-date satellite images of the United States, as well as atmospheric temperature and moisture data.
- Earth Observatory (URL: http://earthobservatory.nasa.gov E-mail: eobmail@eodomo.gsfc.nasa.gov). This NASA-sponsored web site provides access to new satellite imagery and scientific information about the Earth's atmosphere, oceans, land areas, energy use, and plant and animal life, with a focus on climate and environmental change.
- The Latest Cool Image from the U.S. Storm Prediction Center site (http://www.spc.noaa.gov/coolimg) posts "interesting, beautiful, educational, and/or unusual radar, satellite, or analytical imagery dealing with any weather subject." Past images have included satellite views of large wildfires, the formation of tornadoes and hurricanes, and plumes from volcanic eruptions.

Government Sites

- The U.S. Department of Energy (DOE, at http://www.doe.gov) encompasses several relevant divisions: the U.S. Energy Information Administration (http://www.eia.doe.gov), which is the statistical agency of the DOE; the U.S. Office of Energy Efficiency and Renewable Energy (http://www.eren.doe.gov), formed to revolutionize approaches to energy efficiency and renewable energy technologies; the U.S. Office of Scientific and Technical Information (http://www.osti.gov), responsible for disseminating scientific and technical information resulting from DOE research and development programs; and the Carbon Dioxide Information Analysis Center (http://cdiac.esd.ornl.gov), the DOE's primary climate-change data and information analysis center.
- The Global Change Data and Information System (http://www.globalchange.gov) provides access to global change–related reports and publications from the Carbon Dioxide Information Analysis Center (CDIAC), the U.S. Energy Information Administration (EIA), the National Center for Atmospheric Research (NCAR), and the National Oceanic and Atmospheric Administration (NOAA).
- The National Institute of Environmental Health Studies (http://www.niehs.nih.gov) seeks to reduce the incidence and impact of human illness caused by environmental factors, including climate change.

- The U.S. Department of Agriculture (USDA) Global Change Program Office (http://www.usda.gov/oce/gcpo/index.htm) coordinates global change programs related to agriculture and forestry, publishes fact sheets about greenhouse gas emissions and carbon dioxide sinks in agriculture and forestry, and holds greenhouse gas reporting workshops.
- The U.S. Department of State's Bureau of Oceans and International Environmental and Scientific Affairs (http://www.state.gov/g/oes) coordinates U.S. international ocean, environmental, and health policies. The web site provides access to the U.S. Global Climate Change Policy fact sheet and the U.S. Record of Action to Address Climate Change Domestically fact sheet.
- The Government of Canada Climate Change web site (http://climatechange.gc.ca) provides introductory information on global warming, action plans to mitigate its effect, and additional resources for students, teachers, and the media.

United Nations Sites

- The Intergovernmental Panel on Climate Change (http://www.ipcc.ch) was founded by the World Meteorological Organization (WMO) and the United Nations Environment Programme (UNEP) to assess the scientific, technical, and socioeconomic factors involved in anthropogenic climate change. The IPCC produced the important 1990, 1995, and 2001 Assessment Reports, which may be downloaded from the web site.
- The United Nations Framework Convention on Climate Change web site (http://www.unfccc.de) presents information on greenhouse gas emissions, development and transfer of technologies, the Kyoto Protocol, and the 2002 World Summit on Sustainable Development.

Academic and Research Sites

- The Canadian Institute for Climate Studies at the University of Victoria (http://www.cics.uvic.ca) was initiated by the Meteorological Service of Canada and the province of British Columbia to investigate climate variability. It manages several climate-related research initiatives and provides climate predictions.
- The Center for International Earth Science Information Network (http://www.ciesin.org), affiliated with Columbia University, provides access to information aimed at helping scientists, policy makers, and the public better understand the relationship between humans and the environment.
- The Center for Sea and Atmosphere Research (http://www-cima.at.fcen.uba.ar), supported by the Research Council for Science and Technology of Argentina (CONICET), is associated with the Department of Atmospheric Sciences of the University of Buenos Aires, Argentina. Research is

focused on numerical weather prediction, climate modeling, regional climate variability, and regional anthropogenic effects on climate.

- The Climate Research Unit at the University of East Anglia in England (http://www.cru.uea.ac.uk) is one of the world's leading institutions concerned with the study of natural and anthropogenic climate change. Research focuses on past climate history and its impact on humanity; the course and causes of climate change during the present century; and prospects for the future.
- The Environmental Change Institute at Oxford University (http://www.eci.ox.ac.uk) was founded in 1991 to conduct issue-driven, policy-relevant research for the management of environmental change. Focus areas include climate change, sustainable development, and energy use.
- The International Research Institute for Climate Predictions (IRI) at Columbia University (http://iri.1deo.columbia.edu) works to find ways to help society deal with climate fluctuations. The institute is involved in prediction research, climate monitoring, and development of climate models.
- The University Corporation for Atmospheric Research (http://www.ucar.edu) enhances the ability of universities to observe and collect data about the atmosphere by supplying up-to-date weather data to universities, training weather forecasters, and helping organize international research efforts. UCAR's National Center for Atmospheric Research (http://www.ncar.ucar.edu) conducts research into climate change, changes in atmospheric composition, Earth-Sun interactions, weather formation and forecasting, and the impacts of these atmospheric phenomena on human societies.

Environmental Groups

- Green House Network (http://www.greenhouse.net) is a nonprofit organization whose sole purpose is to build a grassroots movement to stop global warming. The group's Climate Education Project (CEP) holds training sessions and organizes a network of volunteer speakers to present information about global warming at schools.
- The Heat Is Online (http://www.heatisonline.org) is author Ross Gelbspan's site, which broadcasts a 10-minute multimedia presentation on climate change, tracks developments in climate science, catalogues extreme weather events, documents efforts by the fossil fuel industry to fight mitigation, and offers solutions to the global warming problem.

Skeptic Groups

- The Coalition for Vehicle Choice's "Climate, Cars, and Consumers" page (http://www.vehiclechoice.org/cvcclim) includes links that explore the background of the Kyoto Protocol, scientific and policy debates about global warming, and congressional reactions to the debate, all through the eyes of those who doubt the validity of IPCC conclusions on the issue.

Environmental News Sites

- Econet (http://www.igc.org/econet) provides access to stories and editorials written from an environmentalist and activist perspective.
- The EnviroLink Network (http://www.envirolink.org) is a nonprofit organization that provides access to thousands of online environmental resources.
- The Environment News Service (http://ens-news.com) is a daily international wire service that presents late-breaking environmental news. Issues and events covered include legislation, politics, conferences, lawsuits, international agreements, demonstrations, science and technology, and renewable energy.
- The Environmental News Network (http://www.enn.com) offers timely environmental news, live chats, interactive quizzes, feature stories, and debate forums aimed at educating users about major environmental issues.

Energy Use Sites

- Western Fuels Association (http://www.westernfuels.org) is a nonprofit organization that provides coal for the generation of electricity by consumer-owned utilities in Louisiana and the Great Plains, Rocky Mountain, and Southwest regions of the United States.

BIBLIOGRAPHIC RESOURCES

Although the Internet and World Wide Web provide virtually unlimited resources for the researcher, libraries and bibliographic resources are still vital assets to any research project. A bibliographic resource is any type of index, catalog, or guide that lists books, texts, periodicals, or printed materials containing articles or chapters related to a subject.

LIBRARY CATALOGS

Most public and academic libraries have placed their card catalogs online. This allows the user to access a library's catalog from any Internet connection, even from home. Viewing a library catalog in advance enables the researcher to develop a comprehensive bibliographic resource list and reserve these resources before signing offline, saving time and frustration.

The Library of Congress (http://catalog.loc.gov) is the largest library catalog available. This site provides advice on search techniques, lists of resources, and catalogs of books, periodicals, maps, photographs, and more.

Online catalogs can be searched by author, title, and subject headings, as well as by matching keywords in the title. Thus a title search for "climate" will retrieve all books that have that word somewhere in their title. However, since not all books about the climate may have that word in the title, it is still necessary to use subject headings to get the best results.

General Library of Congress subject headings under which information on global warming can be found include, but are certainly not limited to, the following:

Biological Diversity
Biotic Communities
Carbon Dioxide
Climatic Changes
Crops and Climate
Ecosystem Management
Energy Consumption
Environmental Economics
Fossil Fuels
Global Warming
Greenhouse Effect, Atmospheric
Greenhouse Gases
Human Beings—Effects of Climate on
International Economic Relations
Meteorology, Agricultural
Pollution
Weather
Weather Forecasting

Once the record of a book or other item has been found, it is a good idea to check for additional subject headings and name headings that may have been previously overlooked. These can be used for additional research.

BOOKSTORE CATALOGS

Other valuable resources are online bookstore catalogs such as Amazon.com (http://www.amazon.com) and Barnes & Noble.com (http://www.barnesandnoble.com). These sites not only offer a convenient way to purchase books related to global warming but also provide publisher information, lists of related topics and books, and customer reviews. These features allow online bookstore catalogs to be used as another source for annotated bibliographies.

PERIODICAL DATABASES

Most public libraries subscribe to various database services, such as InfoTrac, which offer detailed indexes of hundreds of current and back-issue periodicals. These databases can perform searches based on titles, authors, subjects, or keywords within the text. Depending on the service, the database can provide a listing of bibliographical information (author, title, pages, periodical name, issue, and date), a synopsis and abstract (a brief description of the article), or the article in its entirety.

Many public and academic libraries now have dial-up or Internet access, allowing these periodical databases to be searched from home, school, or the local cybercafé. The periodical database search can often be found in the library's catalog menu or on its homepage. Sometimes a library membership card may be necessary to access the information available on a library's web site, so always check with the desired library for its specific policies.

Another extensive but somewhat time-consuming option for searching for periodicals is to visit the web site of a specific periodical related to climate change topics. Often the web address is the periodical's name with *.com* (if commercial), *.gov* (if it is a governmental publication), *.edu* (if it is a university published journal), or *.org* (if it is a publication produced by a public organization) added. Some of these publication may have several years of back issues online.

LEGAL RESEARCH

Gathering and understanding legal research can be more difficult than simply reading through bookstore catalogs and bibliographical indexes. Once again, the Internet proves to be extremely useful by offering a variety of user-friendly ways to research laws and court cases without paging through volumes of court cases in legal libraries (to which the public may not have access).

FINDING LAWS

When federal legislation passes, it becomes part of the United States Code, the massive compendium of federal law. The U.S. Code can be searched online at several locations. Perhaps the easiest and most comprehensive is the U.S. Code database compiled by the Office of the Law Revision Counsel (http://uscode.house.gov). Another option is the web site of Cornell University Law School (http://www4.law.cornell.edu/uscode/). The web site of the American Society of International Law (http://www.asil.org) has an Environment and Space link that provides access to information on international, multilateral environmental agreements. In general, the fastest way to retrieve a law is by its title and section citation, but phrases and keywords can also be used.

The codes of many states' laws are also available online. Links to the codes for specifics states can be found on the 'Lectric Law Library (http://www.lectlaw.com/inll/1.htm).

KEEPING UP WITH LEGISLATIVE DEVELOPMENTS

When performing legal research, some pertinent legislation may be pending. Pending legislation can frequently be found by looking at advocacy group sites for both national and state issues.

The Library of Congress's Thomas web site (http://thomas.loc.gov) is a user-friendly interface that has many valuable features for keeping up with legislative developments and legal research. Thomas allows the user to access proposed legislation to each house of congress by entering key terms or a bill number. For example, if the researcher is looking for air quality legislation, either the bill number can be entered or, if that is not known, key words ("air pollution," for example) can be searched, and a listing of relevant legislation will be compiled.

Clicking on the bill number of one of the items found will display a summary of the legislation, the complete text, its current status, any floor actions, and any other information available. If the bill number is known from the beginning, the legislation can be accessed directly by entering it into the search field.

FINDING COURT DECISIONS

Similar to laws, legal decisions are recorded and organized using a uniform system of citations. The basic elements are *Plaintiff* v. *Defendant* followed by the volume number, the court, the report number, and the year in parentheses. Here are two examples to illustrate the naming method.

United States Department of Energy v. Ohio, 503 U.S. 607 (1992) In this example, the parties are the U.S. Department of Energy (plaintiff) and the state of Ohio (defendant). The case can be found in volume 503 of the *United States Supreme Court Reports*, and the case was decided in 1992. (The name of the court is not indicated in Supreme Court decisions.)

Chevron U.S.A., Inc. v. Natural Resources Defense Council, Inc., 467 U.S. 837 (1984) Here the parties are Chevron U.S.A. (plaintiff) and Natural Resources Defense Council (defendant), the case can be found in volume 467 of the *United States Supreme Court Reports*, and the case was decided in 1984.

To locate a decision made by the federal court, the level of the court involved must first be determined: district (lowest level, first stage for most trials), circuit (the main court of appeals), or the Supreme Court. Once this question is answered, the case and the court's ruling can be located on a number of web sites by searching for either the citation or the names of the parties. There are two sites in particular that are useful for calling up cases.

The first, the Legal Information Institute web site (http://supct.law.cornell.edu/supct), contains every Supreme Court decision made since 1990, plus 610 of the best-known and frequently referenced Supreme Court cases. The site also provides several links to other web sites that contain earlier Supreme Court decisions.

The other site, Washlaw Web (http://www.washlaw.edu), maintains a comprehensive database of decisions made at all court levels. In addition, the site has a large list of legal topics and links, making it an excellent resource for any type of legal research.

For more information and tips on researching legal issues, read the "Legal Research FAQ" at http://www.eff.org/pub/legal/law_research.faq. The EFF site also explains advanced research techniques, including "Shepardizing," so called for *Shepard's Case Citations*, which explains how a decision is cited in subsequent cases and whether or not the case was later overturned.

CHAPTER 7

ANNOTATED BIBLIOGRAPHY

Numerous books and articles have been published on global warming in recent years, particularly since congressional hearings brought the issue to the attention of policy makers and the public in 1988. This bibliography lists a representative sample of sources on the subject, ranging from scholarly scientific studies to opinion pieces aimed at the general public. Sources have been selected for usefulness to the general reader, currency, and variety of points of view.

Listings are grouped according to area of focus: General Weather and Climate, IPCC Reports, Background Material, Economic Issues, Ethics, Policy and Politics, Science and Research, and Ozone Destruction and Global Cooling. Within each of these subjects, listings are further divided by type: books, articles, Internet documents, and videos. A large number of listings cover multiple aspects of global warming; most of these are included under the "Background Material" heading.

GENERAL WEATHER AND CLIMATE

BOOKS

Allaby, Michael. *Dangerous Weather: A Dramatic Introduction to the Science of Weather.* 8 vols. New York: Facts On File, 2003. The books in this eight-volume series describe the science behind droughts, smog, floods, and major storm systems such as tornadoes and hurricanes. In addition to basic introductions to meteorology and climatology, each volume offers biographical information about important scientists, accounts of actual dangerous weather situations, and advice about safety precautions when facing dangerous weather situations.

Burroughs, William J. *The Climate Revealed.* Cambridge, England: Cambridge University Press, 1999. This well-illustrated book provides an authoritative introduction to the Earth's climate and weather-related phenomena. Among the subjects covered that are relevant to the study of the greenhouse effect and climate change are oceans and currents, clouds, the impact of volcanic eruptions, sunspot activity, weather satellites, ice ages, ozone, ice cores, glaciers, deforestation, El Niño, albedo and radiative effects, renewable energy, climate modeling, and sea level rise. A large number of high-quality color photographs and diagrams are included, as are a glossary and index.

Dunlop, Storm, and Francis Wilson. *Weather and Forecasting.* New York: Macmillan, 1982. This book is a field guide, containing numerous color photographs that will help students identify cloud types and other weather-related phenomena. Diagrams indicate how weather can be predicted by observing current conditions and explain the physics behind various climatic conditions in different regions of the globe. The book includes instructions on how students can compile their own weather records and forecast likely maximum temperatures, overnight temperatures, and precipitation. Also useful are a checklist of factors to consider when making a forecast, a glossary of terms, and an explanation of symbols used on weather maps.

Freier, George D. *Weather Proverbs: How 600 Proverbs, Sayings and Poems Explain Our Weather.* Tucson, Ariz.: Fisher Books, 1989. Freier provides a fun and informative introduction to meteorology by analyzing the accuracy of hundreds of proverbs used by early farmers, hunters, fishermen, and sailors to predict the weather. The book also includes introductory chapters on relative humidity, motions of the atmosphere, fronts and winds, the scattering of light, and thunderstorms.

Gates, David M. *Climate.* Man and His Environment Series. New York: Harper and Row, 1972. This book offers a nontechnical introduction to local, regional, and global climate studies, as well as analysis of the impact of climate on various aspects of animal life, including behavior, migration, and reproduction. The effects of climate on humans are also explored, from health to agricultural practices. The final chapter discusses the degree to which greenhouse emissions and other forms of pollution threaten to alter the global climate.

Griffiths, John F. *Climate and the Environment: The Atmospheric Impact on Man.* London: Elek Books, 1976. Griffiths describes the factors that create global climate patterns, then explores the influence of climate and weather on humans. This influence is readily apparent in clothing design, construction practices and architecture, agriculture, and leisure activities. The book closes with chapters on how urban landscapes influence climate and how humans have attempted to alter the effects of climate through irrigation, artificial stimulation of rain, and frost protection. Each chapter ends with lists of references and suggested readings.

Landsberg, Helmut E. *The Urban Climate.* International Geophysics Series. New York: Academic Press, 1981. This somewhat technical book details the various ways in which large cities influence local weather systems. Covered topics include pollutant-weather interactions, solar radiation, wind alterations, humidity, clouds, precipitation, and the effects of urban atmosphere on humans. Most important for the study of global warming is the chapter on the urban heat island effect, often cited by greenhouse skeptics as a cause of higher temperature readings. The book contains numerous charts, graphs, and diagrams, as well as author and subject indexes.

Linacre, Edward. *Climate Data and Resources: A Reference and Guide.* New York: Routledge, 1992. Linacre illustrates the techniques used to measure and analyze climate data, revealing the efforts of scientists to improve the state of climate research despite the inexact nature of climatology. This is a highly detailed book and is recommended only for those who are comfortable with the technical aspects of the environmental sciences.

Nash, J. Madeleine. *El Niño: Unlocking the Secrets of the Master Weather-Maker.* New York: Warner, 2002. Nash focuses on researchers and their efforts to study El Niño and its effects on world climate. Much of the book concerns research into past weather patterns to determine whether the increasing frequency and intensity of El Niño is connected to global warming.

National Research Council. *Learning to Predict Climate Variations Associated with El Niño and the Southern Oscillation.* Washington, D.C.: National Academy Press, 1996. This report presents the findings of the decade-long Tropical Oceans and Global Atmosphere (TOGA) Program, which gathered data between 1985 and 1994 to address the challenges of El Niño and Southern Oscillation (ENSO) prediction. The study led to improvements in the understanding of short-term climate fluctuations, as well as improvement in the quality of computer models used to predict climate. The book covers the growth and components of the TOGA Program, an overview of what was learned from the program and how this knowledge can be applied to ENSO prediction, and a look into possibilities for future research. Also included are three appendixes and a list of references.

Stein, Paul. *Forecasting the Climate of the Future.* The Library of Future Weather and Climate Series. New York: Rosen, 2001. This book focuses on the use of computer models of the climate in the prediction of the extent and effects of global warming. Included are color photographs, a glossary, a list of useful print and online resources, and an index.

———. *Storms of the Future.* The Library of Future Weather and Climate Series. New York: Rosen, 2001. In this book, Stein describes how weather observers track hurricanes, tornadoes, severe thunderstorms, and winter storms, and the methods they use to make short-range predictions. He also touches on the idea that global warming may affect the intensity of some storms. The book features color photographs, a glossary, a list of useful print and online resources, and an index.

Internet Documents

National Climate Data Center. "2001 Report." Available online. URL: http://lwf.ncdc.noaa.gov/oa/about/ncdc2001report.pdf. Downloaded on March 3, 2003. This report provides an overview of the evolution of climate observation, as well as the status of current efforts to collect climate data and monitor phenomena such as droughts and typhoons.

IPCC REPORTS

Books

Houghton, John T., et al., eds. *Climate Change 2001: The Scientific Basis.* Cambridge, England: Cambridge University Press, 2001. "Summary for Policymakers" also available online (URL: http://www.ipcc.ch/pub/spm22-01.pdf). This contribution of Working Group I to the Third Assessment Report of the IPCC covers the full range of scientific aspects of climate change. It includes a summary for policy makers and chapters on observed climate variability and change, the carbon cycle, atmospheric chemistry and greenhouse gases, the effects of aerosols, model evaluation, projections of future climate change, changes in sea level, detection of climate change, and more.

McCarthy, James J., et al., eds. *Climate Change 2001: Impacts, Adaptation, and Vulnerability.* Cambridge, England: Cambridge University Press, 2001. "Summary for Policymakers" also available online (URL: http://www.ipcc.ch/pub/wg2SPMfinal.pdf). This contribution of Working Group II to the Third Assessment Report of the IPCC explores the vulnerability of socioeconomic and natural systems to climate change. It includes a summary for policy makers; chapters on hydrology and water sources, coastal zones and marine ecosystems, human health, and more; and regional analyses.

Metz, Bert, et al., eds. *Climate Change 2001: Mitigation.* Cambridge, England: Cambridge University Press, 2001. "Summary for Policymakers" also available online (URL: http://www.ipcc.ch/pub/wg3spm.pdf). This contribution of Working Group III to the Third Assessment Report of the IPCC addresses the issues of mitigation of greenhouse gas emissions, managing biological carbon reservoirs, geoengineering, cost of mitigation, and decision-making frameworks. It includes a summary for policy makers.

———. *Methodological and Technological Issues in Technology Transfer.* Cambridge, England: Cambridge University Press, 2000. "Summary for

Policymakers" also available online (URL: http://www.ipcc.ch/pub/sutt-e.pdf). This special report of the IPCC offers an overview of the development and transfer of environmentally sound technologies between and within countries. It includes a summary for policy makers and chapters on trends in technology transfer, international agreements and legal structures, transportation, industry agriculture, forestry, human health, coastal adaptation, case studies, and more.

Nakicenovic, Nebojsa, and Rob Swart, eds. *Emissions Scenarios.* Cambridge, England: Cambridge University Press, 2000. "Summary for Policymakers" also available online (URL: http://www.ipcc.ch/pub/sres.e.pdf). This special report of Working Group III of the IPCC describes future development scenarios, predicts the amount of greenhouse gas emissions that will result from such developments, and explores possible response strategies. The report includes a summary for policy makers.

Penner, Joyce E., et al. *Aviation and the Global Atmosphere.* Cambridge, England: Cambridge University Press, 1999. (URL: www.ipcc.ch/pub/av(e).pdf). This special report of the IPCC considers the effects on the Earth's atmosphere of all the gases and particles emitted by aircraft. It includes a summary for policy makers and chapters on atmospheric ozone, aerosols and cloudiness, potential climate change from aviation, and regulatory and market-based mitigation measures.

Watson, Robert T., ed. *Climate Change 2001: Synthesis Report.* Cambridge, England: Cambridge University Press, 2002. "Summary for Policymakers" also available online (URL: http://www.ipcc.ch/pub/SYRspm.pdf). This book provides a comprehensive summary of the main points of the three separate volumes of the IPCC's Third Assessment Report: *The Scientific Basis; Impacts, Adaptation, and Vulnerability; and Mitigation.* It includes a summary for policy makers and a glossary of terms.

Watson, Robert T., et al., eds. *Land Use, Land-Use Change, and Forestry.* Cambridge, England: Cambridge University Press, 2000. (URL: www.ipcc.ch/pub/srlulucf-e.pdf). This special report of the IPCC examines the scientific and technical implications of carbon sequestrations and the global carbon cycle. It includes a summary for policy makers.

BACKGROUND MATERIAL

BOOKS

Abrahamson, Dean Edwin, ed. *The Challenge of Global Warming.* Washington, D.C.: Island Press, 1989. In this early report on climate change, published by Island Press in conjunction with the Natural Resources Defense Council, various scientists and environmental policy experts introduce the concepts of global warming, the greenhouse effect, acid rain, and ozone depletion; explore the effects of global warming on biological diversity, soil moisture levels, water supplies, sea level rise, and other physical systems; and discuss possible policy responses to these potential problems. Researchers should keep in mind that this book, although full of useful introductory information, is outdated, and its use should be supplemented by data from more recent studies.

Athanasiou, Tom, and Paul N. Baer. *Dead Heat: Globalization and Global Warming.* New York: Seven Stories Press, 2002. The authors explain the scientific basis of global warming theory, outline the political reasons whereby, they believe, governments have not acted to reverse climate change, and argue that both environmental and economic factors must be considered to create a viable solution to the problem.

Bates, Albert K. *Climate in Crisis: The Greenhouse Effect and What We Can Do.* Summertown, Tenn.: The Book Publishing Company, 1990. Part One provides an overview of the theory and possible consequences of global warming. Part Two, which covers solutions to the global warming problem, introduces the philosophy of deep ecology and suggests 21 ways to abate future environmental problems. These include controlling population growth by providing unrestricted free access to birth control, phasing out nuclear energy, encouraging energy efficiency, taxing greenhouse gases, promoting sustainable agriculture, reforesting available land, and reducing overconsumption. The book includes a foreword by then-senator Al Gore.

Benarde, Melvin A. *Global Warning . . . Global Warming.* New York: John Wiley and Sons, 1992. Written on the eve of the 1992 Earth Summit in Brazil, this textbook intended for undergraduates contains detailed chapters covering the Earth's climate and the greenhouse effect, the models used to predict climate change, possible positive and negative effects of global warming, and suggested solutions to the energy problem. Though most of the information remains relevant, many facts and figures should be supplemented with more recent research. The book includes illustrations, diagrams, color plates, and a glossary.

Berger, John J. *Beating the Heat: Why and How We Must Combat Global Warming.* Berkeley, Calif.: Berkeley Hills Books, 2000. Ecology Ph.D. Berger uses nontechnical language to explore the consequences of climate change, rebut the arguments of global warming skeptics, and present practical and affordable solutions that individuals can adopt immediately to help preserve the health of the planet.

Bernard, Harold W., Jr. *The Greenhouse Effect.* Cambridge, Mass.: Ballanger, 1980. Predating the formation of the Intergovernmental Panel on Climate change by eight years, this book may very well stand, as the text on the dust jacket claims, as the first account of global warming aimed at the public. Although much of the data has been superseded by more recent research, of particular interest is the in-depth study of the effects, on various regions of the United States, of the heat waves of the 1930s, when the jet stream reached its maximum northward extent. The record temperature extremes, precipitation contrasts, and subsequent crop failures may, Bernard suggests, be a small sample of what could be expected in a world of rising global temperatures.

Burroughs, William J. *Does the Weather Really Matter? The Social Implications of Climate Change.* Cambridge, England: Cambridge University Press, 1997. Atmospheric physicist and science writer Burroughs combines history, politics, economics, meteorology, and climatology to study the impact of weather extremes on society. The first chapters focus on the possible role of weather in the fall of several ancient civilizations, the coming of the dark ages, the famines and plagues of the 14th century, the agricultural crises of the 19th century, and the Dust Bowl of the 1930s. Subsequent chapters explore how unusually cold winters, storms, floods, and droughts affected historic events during the 20th century. In the second half of the book, Burroughs delves into the implications of the 20th century's global warming trend, the possibility that extreme weather events are increasing in both frequency and severity, and the efforts of scientists to predict future climate trends. Included are graphs, some black-and-white photographs, and a list of references.

Carpenter, Clive. *The Changing World of Weather.* New York: Facts On File, 1991. Carpenter's overview of the climate science emphasizes unusual weather conditions, the reasons why climate changes over time, and the factors involved in global warming. Abnormal weather conditions during the late 1980s included drought in Africa and Europe, flooding in Australia, greatly diminished snowfall in the Alps, gale-force winds along coastal France and England, and some of the warmest global average temperatures ever recorded. Carpenter begins with case studies of these phenomena, then moves on to chapters about past climatic epochs and the science of paleoclimatology, the greenhouse effect, the destruction of the planet's rain forests, and the possible consequences of living in a warmer world. Lavish illustrations and a glossary complement the primary text.

Christianson, Gale E. *Greenhouse: The 200-Year Story of Global Warming.* New York: Walker and Company, 1999. Written more as an interesting historic narrative than a scientific tract, *Greenhouse* traces the emergence of global warming as a concern for humankind, from French scientist Jean-Baptiste-Joseph Fourier's 19th century hypothesis that originated the idea of global warming to the 1997 Kyoto Protocol. Along the way, Christianson follows the development of carbon dioxide–emitting industry, as well as scientific theories that have sought to explain the human impact on the planet's weather patterns. The book includes photographs, an index, and an extensive bibliography.

Dauncey, Guy, and Patrick Mazza. *Stormy Weather: 101 Solutions to Global Climate Change.* Gabriola Island, B.C., Canada: New Society Publishers, 2001. This book suggests strategies that individuals, citizens' organizations, businesses, automobile companies, energy companies, and city, state, and national governments can use to help slow the warming of the global climate. Most of the suggestions center on the concept of reducing reliance on fossil fuels and using renewable energy sources whenever possible. Individuals, for example, are urged to use public transportation, choose energy-efficient appliances, buy green power (energy generated from renewable sources), switch to a vegetarian diet, and invest in solar funds. The extensive introduction includes a timeline and discusses oil exploration, the greenhouse effect, energy data, and sustainable energy.

Dotto, Lydia. *Storm Warning: Gambling with the Climate of Our Planet.* Toronto, Ontario, Canada: Doubleday Canada, 2000. Canadian journalist and science writer Dotto believes strong evidence exists that human-induced climate change is behind the extreme weather conditions of the 1980s and 1990s, and that the worst is yet to come. Disappointed in the lack of progress on the issue despite efforts at climate conferences like Rio in 1992 and Kyoto in 1997, she urges immediate and drastic action to avoid the dire consequences of climate change. Included in the book are a list of acronyms used by organizations (accompanied by their web site addresses), chapter references, a selected bibliography, and an index.

Engelman, Robert. *Stabilizing the Atmosphere: Population, Consumption and Greenhouse Gases.* Washington, D.C.: Population Action International, 1994. The nonprofit Population Action International conducts research into and advocates worldwide access to voluntary, high-quality family planning and health services. In this brief report, the organization seeks to answer questions concerning the impact that population growth has on global warming. Clearly written text and easily understood graphics are used to describe the greenhouse effect and trace the parallels between the rise in atmospheric carbon dioxide levels and the rapid rise of global and local population during the 20th century. Of particular interest is a two-page graph that ranks 126 countries according to their per-capita carbon dioxide emissions in 1990, the highest being the United Arab Emirates (33.11 metric tons per person) and the lowest being Nepal (0.03 metric ton per person). Fully one-half of the nations are shown to be above the sustainable emission level (as of 1990) of 1.69 metric tons or less per person. Following this revelation, the author considers global strategies for stabilizing carbon dioxide levels in the atmosphere.

Fisher, David E. *Fire and Ice: The Greenhouse Effect, Ozone Depletion and Nuclear Winter.* New York: Harper and Row, 1990. Written in a reader-friendly, journalistic style, this book analyzes three interconnected problems that threaten to destroy or permanently damage the Earth's atmosphere. Fisher begins with a discussion of natural atmospheric fluctuations in the distant past, then moves on to examine ways that human activities now affect the Earth. Along the way, he summarizes scientific data and models that illustrate the structure and circulation of the atmosphere and that help researchers understand the mechanics of global climate. The focus of *Fire and Ice*, however, lies in the analysis of how the problems of global warming, ozone depletion, and nuclear winter arose; actions that can be taken to avoid worst-case scenarios; and the consequences of inaction. The book concludes with a brief list of relevant organizations, notes, a bibliography, and an index.

Gallant, Roy A. *Earth's Changing Climate.* New York: Four Winds Press, 1979. Gallant considers both the possibility that the Earth's temperatures will decrease (which, he contends, could lead to mass starvation on a global scale) and the possibility that temperatures will rise (in which case humans would face melting polar ice caps and rising sea levels that would submerge New York, Tokyo, and other coastal cities). Chapters are dedicated to the climates of the past, the work of paleoclimatologists who study the evolution of climate in order to better predict the future, and the causes of climate change, both natural (sunspot activity, cosmic dust clouds, variations in the Earth's orbit and magnetic field) and anthropogenic (industrial heat and pollution, chlorofluorocarbons [CFCs], greenhouse emissions). The text is interspersed with photographs, maps, diagrams, charts, and tables, and ends with a glossary and an index.

Gelbspan, Ross. *The Heat Is On: The High Stakes Battle Over Earth's Threatened Climate.* Reading, Mass.: Addison-Wesley, 1997. In this somewhat alarmist take on the climate change issue, Pulitzer Prize–winning journalist Gelbspan exposes what he views as an ongoing collusion between skeptical scientists and the oil and coal companies that allegedly fund them. These exposés are interspersed with examples of extreme weather phenomena intended to prove the point that global warming is real. The book also includes notes and a lengthy appendix that provides specific scientific rebuttals to points made by global warming skeptics.

Grady, Wayne. *The Quiet Limit of the World: A Journey to the North Pole to Investigate Global Warming.* Toronto, Ontario, Canada: Macfarlane Walter & Ross, 1998. Grady chronicles his trip to the Arctic with a group of scientists researching the effects of global warming. In addition to writing a real-life adventure story complete with characters and suspense, Grady also examines the history of Arctic exploration and provides an overview of global warming.

Gribbin, John. *Future Weather and the Greenhouse Effect.* New York: Delacorte Press/Eleanor Friede, 1982. This early book on global warming introduces the concept of the greenhouse effect and offers insight into the implications of its enhancement by human activity.

———. *Hothouse Earth: The Greenhouse Effect and Gaia.* New York: Grove Weidenfeld, 1990. Gribbin, an award-winning science writer, considers climate change to be the most important challenge facing humans during the first several decades of the 21st century. In *Hothouse Earth*, he examines the scientific basis of the greenhouse effect against the broader subject of natural climatic processes and contemplates the consequences of a warmer planet. As a counterpoint to the scientific rigor of the text, Gribbin evokes the controversial concept of Gaia, the idea that Earth is a living organism that regulates the environment (including temperature) in order to sustain life on the planet. Gribbin worries that the burning of fossil fuels is interfering with the natural carbon cycles that have been a part of the workings of Gaia for millions of years, and may lead to environmental imbalances that are beyond the power of the planetary organism to heal. Diagrams and charts help clarify complex material, and an annotated bibliography provides direction for additional research.

———. *Our Changing Climate.* London: Faber and Faber, 1975. This brief book focuses on natural climate change through the ages, with chapters on historical weather, scientific studies of natural records, climatic patterns and weather predictions, atmospheric circulation, planetary alignments and solar activity, and the costs of climate change.

Gupta, Joyeeta. *Our Simmering Planet: What to Do About Global Warming?* Global Issues Series. London: Zed Books, 2002. Writing for a general audience, Gupta, a senior researcher at the Institute for Environmental Studies in Amsterdam, summarizes scientific evidence that supports the theory that global warming is caused by anthropogenic emissions of greenhouse gases, then examines the likely impact of increases in average temperatures, rising oceans, and shifts in rainfall patterns around the world. She also analyzes the politics of global warming treaty negotiations, including efforts between developed countries and developing countries to agree on a course of action.

Haley, James, ed. *Global Warming: Opposing Viewpoints.* Opposing Viewpoints Series. San Diego: Greenhaven Press, 2002. This book is an excellent starting point for researchers who want to understand the arguments posed by various factions in the ongoing global warming debate. As the title suggests, the book consists of numerous short essays that present opposing points of view in answer to such questions as: Does global warming pose a serious threat? What causes global warming? What will be the effects of global warming? Should measures be taken to combat global warming? Contributors include the Union of Concerned Scientists, author Ross Gelbspan, University of Virginia environmental studies professor S. Fred Singer, and *Wall Street Journal* columnist George Melloan. Each viewpoint is accompanied by a series of questions for the reader to consider, and each chapter ends with a periodical bibliography. At the end of the book readers will find questions for further discussion, a list of organizations to contact, a bibliography, and an index. Illustrations, charts, and maps are interspersed throughout the text.

International Institute for Applied Systems Analysis. *Life on a Warmer Earth: Possible Climatic Consequences of Man-Made Global Warming.* Laxenburg, Austria: International Institute for Applied Systems Analysis, 1981. This brief report, an early overview of the now-familiar global warming scenario, begins with an introduction to the global climate system, then goes on to examine how higher temperatures may affect life on planet Earth.

Johansen, Bruce E. *The Global Warming Desk Reference.* Westport, Conn.: Greenwood, 2001. This college-level collection of information on global warming surveys scientific consensus on the issues, describes recent findings, and considers the arguments of skeptics. Headings include ice melt, warming seas, flora and fauna, human health, projections for the year 2100, and possible solutions.

Kraljic, Matthew A. *The Greenhouse Effect.* The Reference Shelf Series. New York: H. W. Wilson, 1992. This book consists of a collection of articles reprinted from newspapers and magazines, including the *New York Times, Smithsonian, Newsweek, USA Today, Popular Science,* and *Environment.* The articles are divided into sections that address whether global warming constitutes a serious threat to the planet, the measurement and causes of the greenhouse effect, the consequences and costs of doing nothing to address the problem, possible solutions, and policy considerations and responses. Included is an annotated bibliography of articles.

Lamb, H. H. *Climate, History and the Modern World.* London: Methuen, 1982. In the first section of this book, Lamb introduces the causes and effects of climate variation, discussing along the way the influence of humans. In the second section, the author covers 10,000 years of human history against the background of the Earth's climatic history, examining how changes in long-term weather conditions have affected the development of civilization. The third and final section reviews the implications of climate change for contemporary society and the future. This book is intended for the nonspecialist and includes plenty of diagrams, maps, and photographs that help the reader understand the concepts discussed in the text.

Leggett, Jeremy. *The Carbon War: Global Warming and the End of the Oil Era.* New York: Routledge, 1999. Part memoir and part rallying cry for the importance of ensuring the viability of sustainable energy, *The Carbon War* presents a pro-environment argument from a former teacher at the Royal School of Mines and consultant to petroleum companies. After working in these capacities for 10 years, Leggett eventually found the growing evidence of global warming impossible to ignore and switched his allegiance to Greenpeace, where he became director of its Climate Campaign and its Solar Initiative. He later cofounded Solar Century (the United Kingdom's first solar electric power company) and became a distinguished scientist at Oxford University. In *The Carbon War,* Leggett recounts, in chronological order, his lobbying efforts on behalf of Greenpeace between 1989 and the Kyoto meetings in 1997 as he travels the globe from one environmental conference to the next. Along the way, he tells of his occasional triumphs but mostly of his frustration with lawmakers and oil company representatives who refuse to believe that global warming poses a serious threat to the planet.

———. *Global Warming: The Greenpeace Report.* Oxford, England: Oxford University Press, 1990. This collection of essays presents the views of the environmental organization Greenpeace on the science and impacts of global warming, with special emphasis given to policy responses. Reflecting the wide range of topics covered, the diverse group of contributors includes professors and researchers in the fields of climatology, biology, physics, forestry, botany, public health, and energy systems. The book contains many diagrams and charts as well as an extensive notes section.

Lomborg, Bjørn. *The Skeptical Environmentalist: Measuring the Real State of the World.* Cambridge, England: Cambridge University Press, 2001. Lomborg, a former member of Greenpeace, surveys prominent environmental issues with the purpose of proving that many environmental groups have overstated the dangers posed by these problems. In the extensive chapter on global warming, he analyzes scientific data used by the IPCC and concludes that the dangers of global warming, although they exist, have been exaggerated and that estimates of temperature and sea level rise are too high. The cost of mitigating the problem, writes Lomborg, must not exceed the cost of the problem itself. He suggests that short-term emission caps as stipulated by the Kyoto Protocol are not the best solution; rather, governments should invest much more money in research and development of renewable energy sources and ensuring the economic growth of developing nations. In reaching this conclusion, Lomborg puts his faith in the World Trade Organization rather than the IPCC. Extensive charts, graphs, notes, and bibliography are included.

Lyman, Francesca, et al. *The Greenhouse Trap: What We're Doing to the Atmosphere and How We Can Slow Global Warming.* Boston: Beacon Press, 1990. Published under the auspices of the World Resources Institute, a Washington, D.C.–based center for policy research on natural resources and the environment, this book introduces scientific concepts relevant to the study of climate change, examines the problems caused by widespread use of fossil fuels, and presents strategies that policy makers and individuals can use to help stabilize the climate. Included are lists of selected readings, organizations, and resources for energy conservation, gardening, tree planting, and recycling.

Maslin, Mark. *Global Warming: Causes, Effects, and the Future.* Stillwater, Minn.: Voyageur Press, 2002. Maslin provides an overview of global warming, presenting evidence that it poses a real threat to the planet and examining the devastating effects rising temperatures will have on health, agriculture, water resources, coastal regions, storm severity, forests, and wildlife. He also discusses developments in the international politics of global warming up to and including the IPCC Bonn meeting in July 2001.

Mitchell, George J. *World on Fire: Saving an Endangered Planet.* New York: Charles Scribner's Sons, 1991. At the time he wrote this book, Mitchell held the position as Senate majority leader. He had been instrumental in the drafting and passage of the Clean Air Act, as well as many other environmental laws. *World on Fire,* intended for the general reader, examines how global warming, acid rain, depletion of the ozone layer, and the widespread destruction of rain forests have brought the planet to the edge of disaster. Mitchell then suggests both scientific and political means to end these environmental threats on a national and international scale. The book includes a list of sources and an index.

Nishioka, Shuzo, and Hideo Harasawa, eds. *Global Warming: The Potential Impact on Japan.* New York: Springer Verlag, 1999. More than 30 prominent researchers present scientific and technical information on the possible impact of global warming on Japan. Topics include natural ecosystems; agriculture, forestry, and fisheries; water resources and environments; the infrastructure and socioeconomic system; and human health.

Parsons, Michael L., and S. Fred Singer. *Global Warming: The Truth Behind the Myth.* Cambridge, Mass.: Perseus, 1995. The authors discuss the relationship between global climate and the greenhouse effect from a skeptical perspective, pointing out flaws in computer models used to predict climate change, offering insights into the views of the models' creators, and exploring nonanthropogenic factors involved in climate change, such as El Niño and ocean currents.

Pearce, Fred, and John Gribbin. *Essential Science: Global Warming.* Essential Science Series. New York: DK Publishing, 2002. Using a mix of colorful graphics, artwork, and photographs, this introductory book examines the causes and effects of global warming, as well as what can be done to remedy the situation.

Schneider, Stephen H. *Global Warming: Are We Entering the Greenhouse Century?* New York: Vintage Books, 1990. Climatologist and environmentalist Schneider presents his views on the science, history, and policy issues of major climatic changes. He explains why the climate is changing, explores the potential consequences of a continued rise in the use of fossil fuels, and shows what individuals and governments can do to mitigate the effects of global warming.

Singer Fred S. *Hot Talk Cold Science: Global Warming's Unfinished Debate.* Oakland, Calif.: The Independent Institute, 1997. Although poorly organized and extremely repetitive, this book provides an overview of the case against global warming "hysteria" and global climate treaties. Singer, professor emeritus of environmental science at the University of Virginia, combines scientific research with conservative public policy opinions. Includes notes, a list of references, numerous charts and diagrams, and an index.

Stevens, William K. *The Change in the Weather: People, Weather, and the Science of Climate.* New York: Delta, 1999. *New York Times* science reporter Stevens recounts the history of climate change on Earth, emphasizing the ways scientists believe that ongoing cycles of cooling and warming have affected human evolution. This history, however, merely sets the stage for a discussion on the current global warming debate, in which Stevens recounts the evidence for, the impact and consequences of, and the political response to climate change in the 20th century. He also includes a chapter on what he calls the "contrarians," those who are skeptical of the data offered by the Intergovernmental Panel on Climate Change (IPCC). The book includes a list of sources, a selected bibliography, and an index.

Turekian, Karl K. *Global Environmental Change: Past, Present, and Future.* Upper Saddle River, N.J.: Prentice Hall, 1996. Turekian offers a chronological narrative of global environmental change, both naturally occurring and human-induced. Much attention is given to atmospheric changes over time, including global warming, acid rain, and ozone depletion.

Williams, Jerry L. *The Rise and Decline of Public Interest in Global Warming: Toward a Pragmatic Conception of Environmental Problems.* Hauppauge, N.Y.: Nova Science, 2001. Sociologist Williams explores the question of why, even in the face of growing evidence that global warming poses a very real threat to human social systems, global warming has received relatively little media coverage. His technical analysis, based on sociological methods, explores public and media interest in global warming, common sense and scientific interpretations of environmental problems, the nature and origin of global warming theory, and patterns of social complacency that allow the public to willfully maintain its ignorance of potentially catastrophic topics.

Wittwer, Sylvan H. *Food, Climate, and Carbon Dioxide: The Global Environment and World Food Production.* Boca Raton, Fla.: Lewis, 1995. Wittwer discusses the effects of rising levels of atmospheric carbon dioxide on crop production and plant growth around the world, with an emphasis on important food crops whose failure could lead to widespread starvation. Topics addressed include the climate as a resource in food production, potential impacts of rising levels of atmospheric carbon dioxide on crops, food security, and future possibilities for research.

Worldwatch Institute. *State of the World 2002: A Worldwatch Institute Report on Progress Toward a Sustainable Society.* New York: W. W. Norton and Company, 2002. This edition of Worldwatch Institute's annual State of the World series evaluates, in preparation for the 2002 World Summit on Sustainable Development, the progress that has been made in various environmental arenas since the 1992 Earth Summit in Brazil. The chapter on global warming covers the evolving science of climate change research, the development of climate change negotiations since 1992, and the international struggle to implement binding carbon dioxide emission caps despite the U.S. government's refusal to ratify the Kyoto Protocol. The authors urge world leaders to bring the Kyoto Protocol into force as quickly as possible and to reaffirm the IPCC as the world's foremost authority on climate change. The report includes numerous tables, figures, and sidebars.

ARTICLES

Beckerman, Wilfred, and Jesse Malkin. "How Much Does Global Warming Matter?" *Public Interest*, vol. 114, Winter 1994, pp. 3ff. This article considers the problem of climate change and explores the possible effects of global warming. The authors complain that melodramatic environmental problems are afforded more attention by the media, environmentalist pressure groups, and the public. Largely ignored are more mundane but more pressing needs like better sewage systems in the developing world.

Bessieres, Michel. "Global Warming: Ignorance Is Not Bliss." *The UNESCO Courier*, vol. 54, June 2001, pp. 10–12. As the 2001 international climate conference in Bonn, Germany, approached, Bessieres examined the disparity between widespread belief among scientists that global warming poses a significant problem and the general unwillingness of the public and political leaders to take meaningful action to deal with the crisis.

"Bleaching Could Help Corals Survive." *USA Today*, vol. 130, June 2002, p. 11. A study by the Wildlife Conservation Society (WCS) questions the conventional wisdom that bleaching is universally detrimental to coral reefs, instead suggesting that the process is a survival strategy that allows corals to rid themselves of dangerous algae.

Boukhari, Sophie. "Forests: A Hot Deal for a Cooler World." *The UNESCO Courier*, vol. 52, December 1999, pp. 10–13. Boukhari explores the use of trees as carbon sinks, a means to absorb and store carbon dioxide and thus reduce the concentration of greenhouse gases in the atmosphere. Some industrialists have used the idea to argue that as long as more trees are planted, curbing the use of fossil fuels is not necessary to fight global warming.

Bowers, Clare L. "Global Warming Shrinks Fish Habitat." *Environment*, vol. 44, October 2002, p. 8. A study by Defenders of Wildlife and the Natural Resources Defense Council claims that greenhouse gas emissions could contribute to the eventual disappearance trout and salmon—coldwater fish that are "already living at the margin of their tolerance"—from U.S. waterways. According to the study, a continuing rise in global temperatures could cause habitat for these species to shrink by as much as 17 percent by 2030, 34 percent by 2060, and 42 percent by 2090.

Bransford, Kent J., and Janet A. Lai. "Global Climate Change and Air Pollution: Common Origins with Common Solutions." *The Journal of the American Medical Association*, vol. 287, May 1, 2002, p. 2,285. In the United States, 98 percent of human-caused carbon dioxide emissions results from fossil fuel combustion. Although this increased level of carbon dioxide has been connected to global warming, it has not been proven to directly affect human health. The combustion of fossil fuels, however, produces airborne particulates, nitrogen oxides, sulfur oxides, and ground-level ozone, all of which are known to have negative impacts on respiratory health. The authors, stressing the idea that fossil fuel combustion is linked to both global warming and air pollution, suggest both problems can be solved simultaneously by developing technologies that increase energy efficiency and by making the transition from fossil fuels to renewable energy resources.

Brown, Kathryn S. "Taking Global Warming to the People." *Science*, vol. 283, March 5, 1999, pp. 1,440–1,441. Growing evidence of the causes and effects of anthropogenic climate change has convinced many scientists that a strong case must be made to deal with global warming now rather than later.

Brown, Stuart F. "How Do You Feel About Nuclear Power Now?" *Fortune*, vol. 145, March 4, 2002, pp. 130–134. Brown reports on the revival of interest in nuclear power due to increasing energy demands and worries that combustion of fossil fuels may cause global warming. In response, the Nuclear Regulatory Commission is debating strategies to protect nuclear power plants and their by-products from infiltration by terrorists.

Denecke, Christl. "Corals Under Siege." *The UNESCO Courier*, vol. 54, March 2001, pp. 10–11. Denecke examines the massive coral die-offs, thought to result from pollution, overfishing, and global warming. The destruction threatens the health of coastal communities that rely on coral reefs to attract tourists, develop the capacity for commercial fishing, and protect shorelines from erosion and damage.

De Roy, Tui. "Caught in a Melting World." *International Wildlife*, vol. 30, November/December 2000, pp. 12–19. The author examines the role of Antarctic Adélie penguins as the "canaries in the coal mines" of global warming because their food supplies are closely tied to patterns and movement of sea ice. Changes in those patterns have already resulted in drastic population declines in several penguin rookeries.

Easterbrook, Gregg. "Greenhouse Common Sense." *U.S. News & World Report*, vol. 123, December 1, 1997, pp. 58–62. This article analyzes the state of the global warming debate on the eve of the 1997 United Nations climate change summit in Kyoto, Japan.

Edmondson, Brad. "What If . . . the Oceans Rose Three Feet?" *American Demographics*, vol. 19, December 1997, p. 44. Edmondson explores the possible effects of sea level rise ("as much as three feet" in the 21st century, according to the IPCC) on coastal populations. Low-lying countries that depend on small farms for food, such as Bangladesh, could suffer from famine. In the United States, about four in 10 people live within 80 kilometers (50 miles) of a coastline. Such coastal cities as New York, Miami, and New Orleans would face inundation or costly engineering projects to avoid flooding, while land that is now safe from ocean storm surges would become vulnerable flood zones.

"Engineer Heads U.N. Climate Panel." *IIE Solutions*, vol. 34, June 2002, p. 14. This brief article profiles Rajandra K. Pachauri, who replaced Robert Watson as chair of the United Nations Intergovernmental Panel on Climate Change under controversial circumstances.

Flynn, Julia, Heidi Dawley, and Naomi Freundlich. "Green Warrior in Gray Flannel: Why Business Listens to Activist Jeremy Leggett." *Business Week*, May 6, 1996, p. 96. The authors profile Jeremy Leggett, who at the time of this article was the director of Greenpeace International's solar campaign. He has found success in convincing insurance and bank executives that global warming could bankrupt their industries.

Foley, Grover. "The Looming Environmental Refugee Crisis." *The Ecologist*, vol. 29, March/April 1999, pp. 96–97. Foley examines the possibility that global climate change will render entire regions uninhabitable through food and water shortages and will inundate towns and cities with water from rising sea levels and increased flooding. The resulting devastation could create millions of new environmental refugees.

Frey, Darcy. "George Divoky's Planet." *New York Times Magazine,* January 6, 2002, pp. 6ff. Scientist George Divoky spends each summer on Cooper Island, a remote barrier island off the northern coast of Alaska, analyzing the connections between a colony of Arctic seabirds and global warming. Frey provides a firsthand account of her adventurous visit to the island to share in Divoky's tribulations and observe his research methods. The scientist had been studying the birds before he realized that from 1975 to 1995 snow in northern Alaska was melting five days earlier each decade, which corresponded to the observation that the birds had laid their eggs an average of five days earlier each decade over the same 20 years.

Galtie, Alain-Claude. "Is El Niño Now a Man-Made Phenomenon?" *The Ecologist,* vol. 29, March/April 1999, pp. 64–67. This article explores the question of whether global warming has contributed to the increasing intensity and frequency of El Niño, which has severely disrupted agriculture and the economies of affected nations.

George, Alan. "Mediterranean Faces Climatic Catastrophe." *Middle East,* January 1998, pp. 17–18. George writes about a report issued by Greenpeace warning that nations in North Africa and the Middle East are likely to be among those hardest hit by global warming. Among the problems, according to the international environmental group, will be acute water shortages, desertification, and flooding of coastal areas such as the Nile Delta from sea level rise.

"Global Warming Triggers Glacial Lakes Flood Threat." *U.N. Chronicle,* vol. 39, September–November 2002, pp. 48–49. This article discusses a report by the United Nations Environment Programme (UNEP) and the International Centre for Integrated Mountain Development (ICIMOD) that found at least 44 new glacial lakes in the Himalayas resulting from accelerated melting of glaciers. Scientists involved in the study worry that the lakes are filling so rapidly they could burst their banks in only a few years, sending millions of gallons of floodwaters down valleys and putting tens of thousands of lives at risk.

Goch, Lynna. "It's Getting Hot Down Here." *Best's Review,* vol. 100, August 1999, pp. 47–51. Goch reports on a disagreement among insurers around the world about the risks of global warming.

Gough, David. "The Melting Mountain." *Newsweek,* February 25, 2002, pp. 34ff. Gough reports on the human and environmental impacts of climate change on and around Mount Kilimanjaro in Africa. Changing patterns of precipitation (the mountain's annual rainfall has declined every year since 1984) have caused the glaciers to recede, mountain streams to dry up, and regional rain forests to shrink, forcing local settlers to abandon failed homesteads and buy food instead of grow it.

"Greenhouse Emissions Growth Slowed over Past Decade." *Cost Engineering*, vol. 44, February 2002, p. 43. The American Association of Cost Engineers reports on a NASA study showing that the rate of growth of greenhouse gas emissions has slowed since its peak in 1980. This is due in large part to international agreements that led to reduced use of chlorofluorocarbons and slower growth rates of methane and carbon dioxide emissions.

Guterl, Fred. "The Truth About Global Warming." *Newsweek*, July 23, 2001, p. 44. Guterl provides an overview of the life and beliefs of Richard Lindzen, the Alfred P. Sloan Professor of Meteorology at MIT and a respected voice of dissent against the idea that global warming poses a serious threat to Earth. Although he acknowledges that the Earth warmed slightly during the 20th century and that human activity might be at least partly to blame, Lindzen believes that the IPCC's predictions of how much warming will occur are greatly exaggerated.

Hakim, Danny. "Smokestack Visionary." *New York Times Magazine*, September 29, 2002, Section 6, p. 100. Larry Burns, vice president for research and development and planning at General Motors, is working on the Hy-wire, an ambitious prototype car that runs on a hydrogen fuel cell and features a power system that creates an electrical current from chemical reactions. Although Burns believes there is enough evidence for carmakers to take global warming seriously, his stance is at odds with that of General Motors vice chairman Bob Lutz, a global warming skeptic who talks of the Kyoto Protocol as a conspiracy theory. Environmentalists have viewed Burns's efforts with suspicion, noting that General Motors is responsible for the gas-guzzling Hummer and that the company has argued that being forced to make more efficient cars now will only undermine their work on the fuel cell for the future.

Hansen, Eric. "Climate Change: Himalayan Liquidation." *OnEarth*, vol. 24, Fall 2002, p. 8. Hansen reports on the problem of melting glaciers in Nepal, thought to be caused by global warming. Sites such as Imja in the Khumbu region of Nepal have, in recent years, become dangerously unstable catch basins for millions of gallons of water. Should the icy banks of new lakes like the one at Imja rupture, down-valley farming villages could be wiped out by cascading water and debris. Roger Payne of Union Internationale des Associations D'Alpinisme, mountaineering's international association, noted that while most climbers of the world's major peaks come from nations responsible for the majority of greenhouse gas emissions, it is the poor people of developing countries like Nepal who will suffer most.

Hayden, Thomas, and Masha Gessen. "Killer with a Cold Heart: A Russian Avalanche and a World of Glacial Threats." *U.S. News & World Report*, vol. 133, October 7, 2002, p. 71. This article reports on the September

20, 2002, avalanche that destroyed the village of Karmadon in southern Russia, killing more than 100 people. According to some scientists, global warming may cause such catastrophes to become more common, as melting glaciers on high mountains can break up and slide into valleys below or form giant lakes of meltwater that eventually break through natural dams of ice or debris in cataclysmic "outburst floods."

Heinrichs, Jay. "Camp Apocalypse." *Backpacker,* June 2003, pp. 70–74+. The author gives a firsthand account of a backpacking trip to Alaska's Kenai Peninsula, where an infestation of spruce bark beetles has decimated the forests. He presents evidence that the increased insect population is a direct result of rising temperatures caused by global warming. A companion article, "The Global Warming Tour 2003" by Steve Howe, is a travel guide to eight areas in the United States and Canada where the author claims the environmental effects of global warming can be directly observed. Sidebars provide emissions and temperature statistics, evidence of northward wildlife migrations, and advice on how individuals can act to slow global warming.

Hodgson, Peter, and Yves Marignac. "Debate: Is Nuclear Power a Viable Solution to Climate Change?" *The Ecologist,* vol. 31, September 2001, pp. 20–23. In this point/counterpoint article, Hodgson argues that dismantling power stations that use fossil fuels and replacing them with nuclear power plants is the most realistic way of fighting global warming in the immediate future. Marignac contends that nuclear waste is no more desirable than fossil fuel emissions and that more effort should be directed toward developing renewable energy sources.

Honeywill, Tristan. "Climate Opinion." *Professional Engineering,* vol. 15, January 16, 2002, pp. 31–32. Honeywill takes a close look at James Lovelock's controversial Gaia hypothesis—which holds the view that the Earth is an actively self-regulating superorganism—and its influence on theories about global warming. Lovelock also believes that until renewable energy becomes viable as the primary source of the world's energy, an immediate switch should be made from fossil fuels to nuclear power, as the latter does less damage to the atmosphere. This idea has raised controversy even among those who have come to accept the Gaia hypothesis.

Horton, Tom. "Stormy Weather." *Rolling Stone,* March 20, 1997, pp. 62–70 ff. Horton looks at the work of scientists who are researching the question of whether global warming is to blame for an apparent increase in extreme weather around the world.

Hsiao, Andrew. "Greenwash." *The Village Voice,* vol. 43, October 13, 1998, pp. 26ff. In 1998, Ford Motor Company signed on as the "exclusive sponsor" of *Time* magazine's "Heroes for the Planet," a series of articles profiling eco-activists. The deal both enraged and amused environmentalists, who have long battled auto manufacturers over air pollution and global warming and who saw the sponsorship as an underhanded effort by Ford to appear sensitive to environmental concerns.

"Ice Sheet Melting Key to Sea Level Rise." *USA Today*, vol. 130, June 2002, pp. 7–8. Researchers have linked a sea level rise of 21 meters (70 feet) in less than 500 years to the partial collapse of ice sheets in Antarctica about 14,200 years ago.

"Is Europe About to Freeze?" *USA Today*, vol. 130, June 2002, pp. 8–9. This article examines the strange possibility that global warming could alter ocean current patterns in such a way that much of Europe could suffer abrupt temperature drops and develop a climate resembling that of Alaska.

Jennings, Lane. "Climate Change: What We Can Do." *The Futurist*, vol. 36, January/February 2002, p. 68. Using Guy Dauncey's book *Stormy Weather: 101 Solutions to Global Climate Change* as a guideline, Jennings offers a list of things that individuals, citizens' organizations, businesses, and nations can do to fight global warming. These include traveling more sustainably and making homes more energy efficient (individuals); organizing a car-free Sunday (citizens' organizations); upgrading lighting and insulating roofs and windows (businesses): and ending subsidies for fossil fuels (nations).

Kauffmann, Bruce G. "Is the Gas Always Greener?" *American Gas*, vol. 77, June 1995, p. 30. This article examines the relationship between the natural gas industry and environmentalists. Kauffmann explains that the natural gas industry must frequently remind environmentalists that burning natural gas produces up to 30 percent less carbon dioxide than oil, and up to 45 percent less than coal, making gas a viable energy alternative for the fight against global warming.

Kingsnorth, Paul. "Human Health on the Line." *The Ecologist*, vol. 29, March/April 1999, pp. 92–94. Kingsnorth examines the IPCC's contention that "climate change is likely to have wide-ranging and mostly adverse impacts on human health, with significant loss of life."

Kluger, Jeffrye. "A Climate of Despair." *Time*, vol. 157, April 9, 2001, pp. 30–36. While the rest of the world seems to be moving toward a solution to the global warming problem, environmentalists worry that the Bush administration's unwillingness to take effective action on the issue may derail the entire process.

Kriz, Margaret. "Chilling Out." *National Journal*, vol. 29, May 3, 1997, pp. 866–869. Kriz reports on how many U.S. corporations, after years of resisting action to stop global warming, are beginning to consider the economic benefits of adopting environmentally friendly business practices.

———. "This Corporate Group Ran Out of Gas." *National Journal*, vol. 34, February 9, 2002, p. 407. This brief article notes the disbandment of the Global Climate Coalition (GCC), a conservative organization representing 15 trade associations that spent 12 years fighting against international controls on carbon dioxide emissions and other pollutants. A GCC spokesman claimed the group was "deactivated" because it had succeeded in its attempt to convince the United States to reject the Kyoto Protocol. Environmentalists pointed out, however, that many companies—including General Motors, DaimlerChrysler, Ford, BP Amoco, and Shell Oil—had recently severed ties with the GCC to pursue more moderate stances on global warming.

Kuchment, Anna. "State of the Ice." *Newsweek*, February 25, 2002, pp. 33ff. Although most scientists agree that the Earth is warming, the science of predicting climate change remains uncertain. Kuchment examines several areas of debate on the subject, including the massive ice sheet along the Ross Sea in Antarctica, which is gaining in mass rather than shrinking as had been previously thought. Another important area of disagreement is the rate at which climate change occurs: Conventional wisdom had pointed to a slow, steady change, but the National Academy of Sciences has warned of the possibility of abrupt changes that occur in a decade or two, rather than over centuries.

Lal, Rattan. "A Modest Proposal for the Year 2001: We Can Control Greenhouse Gases and Feed the World . . . with Proper Soil Management." *Journal of Soil and Water Conservation*, vol. 55, Fourth Quarter 2000, pp. 429–433. This article examines the interconnection of the rapid increase in world population, the continuing degradation of agricultural soils, and global warming. The author suggests the agricultural practices that prevent soil degradation can reduce greenhouse gas emissions and ensure a secure global food supply.

Landers, Jay. "Climate Change to Alter California's Water Supplies, Study Says." *Civil Engineering*, vol. 72, August 2002, pp. 16–17. A study by researchers at the University of California at Santa Cruz and the Lawrence Livermore National Laboratory suggests that global warming could cause major shifts in the nature and timing of precipitation in California. This could significantly affect how and when California procures its public water supplies.

LaRochelle, Mark, Peter Spencer, and S. Fred Singer. " 'Global Warming' Science: Fact vs. Fiction." *Consumers' Research Magazine*, vol. 84, July 2001, p. 7. The authors produce a litany of grievances against scientists who say global warming poses a serious threat, the IPCC, and the media, claiming that close scrutiny of climate data reveals that the facts of climate change are actually the opposite of what has been popularly portrayed.

Larson, Vanessa. "New Studies Show Cooling and Ice-Sheet Thickening in Antarctica." *World Watch*, vol. 15, May/June 2002, p. 9. Larson reports that a study conducted by researcher Peter Doran at the University of

Illinois at Chicago shows that temperatures in Antarctica's McMurdo Dry Valleys region dropped by 0.7 degree Celsius (1.3 degrees Fahrenheit) per decade between 1986 and 2000. Previous studies that had indicated rising temperatures in the area had, according to Doran, relied too heavily on temperature data from the Antarctic Peninsula, which "has been warming drastically." Despite this cooling trend, Doran maintains it does not indicate a reversal of global warming on a worldwide scale, which he says "is real and happening right now" even though "the Antarctic region is not behaving as predicted."

Lemonick, Michael D. "How to Prevent a Meltdown." *Time*, vol. 155, Spring 2000, pp. 60–63. Many climate experts agree that the best way to slow the warming of the planet is to move away from dependence on fossil fuels and restructure the way energy is produced. Lemonick discusses alternatives to oil, gas, and coal that would reduce atmospheric pollution.

———. "Life in the Greenhouse." *Time*, vol. 157, April 9, 2001, pp. 24–29. Lemonick profiles MIT atmospheric scientist Richard Lindzen and John Christy of the University of Alabama in Huntsville, two global warming skeptics who agree that humans are influencing the climate but believe that the IPCC is overestimating the extent and potential damage of global temperature rise.

Levine, Mark. "Tuvalu Toodle-oo." *Outside*, December 2002, pp. 90–96ff. Tuvalu is a tiny island nation in the Pacific Ocean that has loudly declared to the international community that it is on the verge of being submerged due to sea level rise caused by global warming. The government has asked both Australia and New Zealand to let citizens of Tuvalu immigrate when the islands disappear, and it has considered suing the United States and Australia for releasing a large percentage of the greenhouse gases that have caused global warming. Levine visited the island expecting to find an island paradise going down because of a "single high-profile cosmic phenomenon." Instead, he found a nation beset by more mundane problems that accompany globalization in general: overflowing landfills, overcrowded conditions, and limited economic prospects.

Lindsay, James M. "Global Warming Heats Up: Uncertainties, Both Scientific and Political, Lie Ahead." *The Brookings Review*, vol. 19, Fall 2001, pp. 26–29. This article analyzes the manner in which uncertainty has shaped the debate over global warming in the United States. Lindsay contends that although U.S. nonparticipation in the Kyoto Protocol will not necessarily render the international agreement impotent, the Bush administration must commit to its own plan to reduce emissions for anything to be done about climate change on a global scale.

Lindzen, Richard S. "Two Case Studies: How Political Advocates Misuse Science." *Consumers, Research Magazine*, vol. 79, September 1996, p. 19.

Lindzen uses global warming and eugenics to illustrate the ways that scientific research into sensitive areas can be misused by policy makers and misunderstood by the public.

Livesey, Sharon M. "Global Warming Wars: Rhetorical and Discourse Analytic Approaches to ExxonMobil's Corporate Public Discourse." *The Journal of Business Communication*, vol. 39, January 2002, pp. 117–148. Livesey utilizes highly technical communications theory to analyze the effectiveness of four advertorials, or issue advocacy advertisements, on climate change published by ExxonMobil in *The New York Times* in March and April 2000. These advertisements were part of the corporation's efforts to convince readers to support U.S. government inaction on the issue of carbon dioxide emission reductions. The paper includes footnotes and a lengthy list of references.

Malin, Clement B. "Petroleum Industry Faces Challenge of Change in Confronting Global Warming." *Oil & Gas Journal*, vol. 98, August 28, 2000, pp. 58–63. Malin, former vice president of international relations at Texaco, discusses ways that the petroleum industry can respond positively to public concern about the threat of global warming. He suggests that oil companies work to reassert their leadership roles in promoting growth in a rapidly expanding global economy.

Mann, Charles C. "Getting over Oil." *Technology Review*, vol. 105, January/February 2002, pp. 32–38. Although environmentalists have lobbied for a transition to renewable energy sources as the best way to combat global warming, relatively low electricity prices mean that most nations and energy companies lack financial incentive to develop cleaner technologies. Mann analyzes the belief among some economists that uncertainty about energy supplies stemming from terrorist attacks and general instability in the Middle East may provide the best incentive for developing alternatives to fossil fuels.

Markels, Alex. "Defrosting the Past: Ancient Human and Animal Remains Are Melting out of Glaciers, a Bounty of a Warming World." *U.S. News & World Report*, vol. 133, September 16, 2002, p. 63. Markels reports on how decades of unusual warmth, thought by many to be linked to carbon dioxide emissions, have shrunk or thawed many of the world's glaciers and revealed a treasure trove of well-preserved human and animal remains. Scientists examining the fabrics, wood, bone, and DNA-rich tissue that have been locked in ice for centuries are gaining new understanding of the health, habits, and technology of ancient humans.

Masibay, Kim. "Does My Gas Cause Global Warming?" *Science World*, vol. 58, January 21, 2002, pp. 18–19. Masibay provides a brief look into the production of methane, a potent greenhouse gas, by livestock. The gas is also produced by petroleum drilling, coal mining, solid waste landfills, rice paddies, and wetlands. Although it is less abundant than carbon dioxide, its heat-trapping power in the atmosphere is 21 times stronger than that of the more common greenhouse gas. The article includes a look at methods being used by farmers around the world to reduce and recycle emissions from livestock.

Matthews, Robert. "Climatic Changes That Make the World Flip." *The UNESCO Courier*, vol. 52, November 1999, pp. 10–13. Matthews reports on research showing that, contrary to conventional wisdom, climate change in the past has often occurred in dramatic, rapid "flips" rather than as gradual changes.

McManus, Reed. "Gullible Warming?" *Sierra*, vol. 83, July/August 1998, p. 66. In April 1998, Frederick Seitz, chair of the Marshall Institute, issued a report discrediting international efforts to deal with global warming. The document was printed in a format and typeface that mimicked the journal of the National Academy of Sciences (NAS). The academy responded by stating that the report "does not reflect the conclusion of expert reports of the academy."

"The Melting Continent." *Environment*, vol. 44, May 2002, pp. 7ff. This brief article reports the March 2002 collapse of the 195-meter-thick (650-foot-thick) Larsen B ice shelf in the Antarctic. Although the rapid melting could not definitely be attributed to global warming, scientists are hard pressed to provide an alternate explanation for the collapse of ice shelves that have existed for thousands of years.

Mencimer, Stephanie. "As the World Burns." *The Washington Monthly*, vol. 34, September 2002, pp. 8–9. Mencimer examines the contradiction between President George W. Bush's unwillingness to confront the problem of global warming and the idea that because of its topography, Texas is second only to California among the states likely to be the hardest hit by climate change, which is expected to increase temperatures and reduce rainfall in the already hot and arid state.

Michaels, Patrick. "Computer Models for Weather Forecasting Are Ridden with Errors." *Insight on the News*, vol. 16, March 6, 2000, p. 46. Global warming skeptic Michaels takes a critical look at computer models of the atmosphere in light of an erroneous forecast in January 2000. Despite predictions by models of rapid global warming, Michaels contends that "independent" measures of temperature have shown no net warming over the past 20 years.

———. "Global Warming Warnings: A Lot of Hot Air." *USA Today*, vol. 129, January 2001, pp. 18–20. Michaels argues that global warming, rather than being an imminent apocalyptic event, is a gradual and benign process that has resulted in longer life spans, better nutrition, and increased wealth.

Mihm, Stephen. "Global-Warming Lawsuits." *New York Times Magazine*, December 9, 2001, Section 6, pp. 76 ff. In the summer of 2001, Andrew Strauss, a law professor at Widener University in Delaware, proposed the idea that nations could file lawsuits against the United States to recover damages from emissions of carbon dioxide. Candidates for litigation are plentiful, including such low-lying regions as the Netherlands, Bangladesh, and any number of South Pacific island nations that could be adversely affected by rising sea levels. The main difficulty, according to Strauss, lies in finding courts that will take such cases.

Mills, Evan. "Insurers Have Much to Gain, Lose in Battle to Ease Threat of Global Warming." *National Underwriter*, vol. 102, September 21, 1998, pp. 33–34ff. Mills urges insurers to become expert advisers and information providers on climate change and energy efficiency to reduce the likelihood of insured losses at the hands of extreme weather events.

Monastersky, Richard. "Acclimating to a Warmer World." *Science News*, vol. 156, August 28, 1999, pp. 136–138. Some scientists who believe that reducing carbon dioxide emissions is only a partial solution to the global warming problem are looking into ways to adapt to a warmer world. The approach, according to Monastersky, does not represent an abandonment of mitigation strategies so much as a broadening of paths.

———. "Satellites Misread Global Temperatures." *Science News*, vol. 154, August 15, 1998, p. 100. This article reports on an error in satellite records of the Earth's temperature arising from changes in the satellites' orbits. The flaw, according to researchers, artificially depressed atmospheric temperature readings and masked signs of increasing air temperatures. Scientists involved in the project contend that the discovery strengthens the case for global warming by explaining the enigma, long held up by global warming skeptics as proof that the Earth's temperatures were not rising, of why satellite data have been at odds with surface measurements of global temperature.

———. "The Storm at the Center of Climate Science." *The Chronicle of Higher Education*, vol. 47, November 10, 2000, pp. A16–A18. This article reports on the fallout from a paper published by James Hansen that many people—including reporters, environmentalists, lobbyists, scientists, and politicians—erroneously interpreted as a renunciation of his views that carbon dioxide pollution must be curbed to stop global warming.

———. "Weather Balloons Deflate Climate Blow-Up." *Science News*, vol. 156, September 4, 1999, p. 150. Monastersky analyzes the implications of measurements made with weather balloons showing that the Earth's surface and the lower atmosphere warmed at different rates during the 1980s and 1990s.

Moore, Curtis A. "Awash in a Rising Sea." *International Wildlife*, vol. 32, January/February 2002, pp. 26–35. Moore provides a well-written firsthand account of the effects of global warming on tropical Pacific islands, where rising sea levels and temperatures have swallowed homes with no warning, washed cemeteries out to sea, turned water supplies brackish and undrinkable, bleached once-pristine coral reefs, disrupted important fisheries, and eroded beaches.

Moore, Thomas Gale. "Why Global Warming Would Be Good for You." *Public Interest*, Winter 1995, p. 83. Moore contends that evidence supporting theories of global warming is shaky. However, if the Earth is warming, the likely rise will be so small that it will go virtually unnoticed by inhabitants of the industrial countries. However, if warming does become apparent in the future, it will likely benefit most of humankind by promoting an expansion of agriculture. If global warming does create more problems than benefits, Moore concedes that policy makers may have to consider measures to undo the damage.

Myers, Norman. "Environmental Refugees in a Globally Warmed World." *Bioscience*, vol. 43, December 1993, pp. 752ff. Myers explores the problems involved with estimating the scope of future population displacement caused by global warming. Bangladesh, Egypt, China, India, and Pacific Ocean island states are used as examples of problems that could result from rising sea levels. Problems with agriculture and large cities are also discussed.

Nash, Madeleine. "Wait Till Next Time." *Time*, vol. 154, September 27, 1999, pp. 38–40. Nash discusses the influence that global warming may have on the formation of Atlantic Ocean hurricanes, which depend on warmer water to gain energy.

Natural Resources Defense Council. "Carbon Kingpins: The Changing Face of the Greenhouse Gas Industries." *Multinational Monitor*, vol. 20, June 1999, pp. 9–14. This report by the Natural Resources Defense Council examines the amount of carbon-based pollution produced by the largest energy companies.

"New Weapons Help Fight Global Warming." *Resource*, vol. 6, July 1999, pp. 3–4. This article reports on a workshop sponsored by the U.S. Department of Energy and the Council for Agricultural Science and Technology. Participants concluded that farms, forests, and grasslands can be used to combat global warming by removing carbon from the atmosphere and storing it in the soil.

"NOAA Reports Record Warmth for January–March 2000." *Bulletin of the American Meteorological Society*, vol. 81, July 2000, pp. 1,623–1,624. Measurements taken by the National Oceanic and Atmospheric Administration's National Climatic Data Center indicate that in the year 2000 the United States experienced the warmest January–March period in the 106 years that data has been collected. The data also show that June 1999–March 2000 was the warmest June–March on record.

"No Consensus on Climate." *Oil and Gas Journal*, vol. 99, August 27, 2001, p. 19. This anonymous article asserts that while the United Nations' Intergovernmental Panel on Climate Change (IPCC) Third Assessment Report represents an important step in the scientific understanding of global warming, it should not be taken as consensus on scientific fact. The author cites global warming skeptic Richard S. Lindzen's claim that the "Summary for Policymakers" that most people will read does not accurately reflect the findings of the full IPCC report.

Palmer, Andrew C. "If Global Warming Is Real, How Should Humankind Respond?" *Oil and Gas Journal*, vol. 99, July 30, 2001, pp. 24–26. Palmer, a research professor of petroleum engineering at Cambridge University, argues that if global warming is real, there is nothing that can be done to stop it. Attempts to replace fossil fuels with renewable energy would be futile, as humanity does not possess the wherewithal to develop such technologies. Palmer also states that humans possess the technical genius to adapt to the inevitable effects of climate change.

Parris, Thomas M. "A Look at Climate Change Skeptics." *Environment*, vol. 40, November 1998, pp. 3ff. Parris profiles critics of carbon dioxide mitigation strategies. He divides them into two camps: those who focus on the science of climate change and those who are concerned with the specifics of the Kyoto Protocol. Among the groups mentioned are the Center for the Study of Carbon Dioxide and Global Change, the Global Climate Information Project, and S. Fred Singer's Science and Environmental Policy Project (SEPP).

Patz, Jonathan A., and Mahmooda Khaliq. "Global Climate Change and Health: Challenges for Future Practitioners." *The Journal of the American Medical Association*, vol. 287, May 1, 2002, pp. 2,283–2,284. Global warming and the resulting heat waves, floods, and droughts are expected to cause increasingly broad and severe health problems for humans. Among the potential challenges faced by health practitioners are increased concentrations of regional air pollutants and allergens; disrupted water and food supplies, which could increase the spread of infectious disease and cause widespread malnutrition; and rising sea levels, which could lead to major population displacement and economic disruption.

Paul, Pamela. "It's Too Darn Hot." *American Demographics*, vol. 23, July 2001, p. 22. Paul reports the results of an August 2000 Harris poll in which most Americans questioned said they are aware of global warming and believe the theory that increased emissions of greenhouse gases will lead to an increase in average temperatures. Far fewer respondents, however, are concerned about specific impacts global warming might have.

Pearce, Fred. "Hot Enough for You?" *The Ecologist*, vol. 32, March 2002, pp. 38–42. Pearce conducts a worldwide survey of erratic weather patterns, unprecedented drought conditions, melting glaciers, sea level rise, disease outbreaks, and, as the Intergovernmental Panel on Climate Change (IPCC) calls them, "heavy and extreme precipitation events." All of these are taken by the author to provide evidence of global warming. The article focuses on the lives of people who are affected by these extreme weather conditions.

Petit, Charles W. "The Great Drying: Blame It on Distant Oceans—and Perhaps on Global Warming." *U.S. News & World Report*, vol. 132, May 20, 2002, pp. 54ff. Petit reports on the widespread drought conditions in the United States in the spring of 2002. Using computer models, scientists have determined that the likely causes of this unusually dry climate are La Niña conditions in the eastern Pacific off the coast of Peru, and a vast stretch of warm tropical water in the Indian Ocean, which has steadily been getting warmer for 50 years, perhaps attributable to global warming.

Pohl, Frederik. "Averting Future Disasters." *The Futurist*, vol. 33, March 1999, pp. 34–40. Pohl. author of many well-known science fiction and nonfiction books, considers a wide variety of possible disasters that might befall the planet in the future. Although there is nothing humans can do about cosmic events, other events, such as terrorism and global warming, could be avoided if policy makers take the right actions now.

Power, Carla. "Hitting a Handy Villain: Activists Boycott Esso over Global Warming." *Newsweek*, May 28, 2001, p. 26. Power reports on a campaign started by Greenpeace and Friends of the Earth in England to boycott Esso's products until its parent company, ExxonMobil, changes its position on global warming. The Stop Esso campaign accuses Exxon of pressuring President George W. Bush to abandon the Kyoto agreement by running advertisements criticizing the accord and by contributing more than $1 million to the Republican Party during the 2000 elections. Campaign spokesman Rob Gueterbock likens Exxon's denial of the link between fossil fuels and global warming to the tobacco industry's denial that smoking is linked to cancer.

"Replacing Gas with a Gas: Hydrogen-Powered Cars." *The Economist*, vol. 360, July 21, 2001, p. 66. Auto manufacturer BMW has unveiled a prototype version of a car that has a hydrogen-powered, internal combustion engine as an alternative to fuel cells, which are still several years from commercial viability in cars. The simple process of converting engines to run on hydrogen would help reduce global warming because hydrogen can be produced from renewable sources and does not release carbon dioxide when it is burned. The article points out two catches to this technology: Fuel cells are more efficient than burning hydrogen, and hydrogen would reduce engine performance, since it contains significantly less energy than gasoline.

Retallack, Simon, and Peter Bunyard. "We're Changing Our Climate! Who Can Doubt It!" *The Ecologist*, vol. 29, March/April 1999, pp. 60–63. The authors contend that the mass media has created the false impression that the question of global human-induced climate change is still open to debate. On the contrary, the authors claim, there is an overwhelming consensus among scientists that human-induced climate change is real and is occurring now.

Revkin, Andrew C. "The Heat Is On." *New York Times Upfront*, vol. 124, April 8, 2002, pp. 14–17. Revkin reports on President George W. Bush's tendency to make policy decisions about global warming according to what is best for the fossil fuel industry rather than what is best for the environment. After rejecting the Kyoto Protocol, Bush announced that he would encourage industry to voluntarily limit, but not halt, increases in greenhouse gas emissions. This plan, according to many experts, almost ensures that the planet faces a much warmer future.

Ridgley, Heidi. "How Global Warming Affects Your Allergies." *National Wildlife*, vol. 40, April/May 2002, p. 34. Ridgley reports on research conducted by Lewis Ziska, a plant ecologist with the U.S. Department of Agriculture (USDA) and an allergy sufferer, on the effects of greater levels of carbon dioxide in pollen-producing plants. He concluded that increasing amounts of the greenhouse gas in the atmosphere may have doubled ragweed's pollen-producing capacity over the last 100 years; if emissions continue unabated, pollen counts could double again by the end of the 21st century.

Rowen, Henry S., and John P Weyant. "Staying Cool About Global Warming." *The National Interest*, Fall 1999, pp. 87–93. The authors, who believe that global warming poses a real threat, discuss the problems of the Kyoto Protocol and why they think it will do little to reduce human-induced climate change.

Scheer, Roddy. "Parks as Lungs." *E: The Environmental Magazine*, vol. 12, November/December 2001, pp. 15–18. Scheer discusses the important role that urban forests, found within city parks across the United States, play in absorbing pollutants and storing carbon dioxide. According to David Nowak, project leader of the U.S. Forest Service's Urban Forest Ecosystem Research Unit, trees in Chicago remove 15 metric tons of carbon monoxide, 84 metric tons of sulfur dioxide, 89 metric tons of nitrogen dioxide, 191 metric tons of ozone, and 212 metric tons of particulates each year. Much of the article focuses on Central Park in New York City.

Schneider, Doug. "On Thin Ice: Alaska Feels the Heat of Climate Warming." *Alaska*, vol. 67, October 2001, pp. 40–45. Although some scientists cited in this article question the conventional wisdom that global warming is caused by fossil fuel emissions, all agree that temperatures in Alaska have risen drastically in recent decades. Schneider documents some of the evidence and considers the economic and environmental impact of warmer temperatures. He found that Inupiat in villages along Alaska's northwestern coast have been forced to move houses that the ocean has threatened to overwhelm. Warmer temperatures have allowed trees along Alaska's North Slope to move north and take root at rates not seen in the past 8,000 years. Melting permafrost has left 10-meter-deep (33-foot-deep) sinkholes in some parts of the state. Several dog mushing races have been canceled or rescheduled because there has not been enough snow on the ground. Warmer temperatures have also made avalanches and drowning deaths attributed to thinning ice more common. These problems, according to the article, are only the beginning, as most researchers agree that Alaska is in for even more warming, which will make winters less severe and summers warmer and wetter.

Schulz, Kathryn. "Global Warming Right Now." *Rolling Stone*, February 20, 2003, pp. 35–37. Schulz surveys the current state of global warming studies, focusing on the dismay felt among climate scientists at the Bush administration's announced plan to study the issue for five more years rather than take significant action to regulate greenhouse gas emissions.

Shute, Nancy, et al. "The Weather Turns Wild: Global Warming Could Cause Droughts, Disease, and Political Upheaval." *U.S. News & World Report*, vol. 130, February 5, 2001, pp. 44–52. This article examines predictions—based on data collected from satellites, weather balloons, and computer models of the global climate system—that global average temperature will rise by 1.4 to 5.8 degrees Celsius (2.5 to 10.4 degrees Fahrenheit) during the 21st century, possibly causing widespread environmental damage.

Sissell, Kara. "World CO_2 Emissions Dip." *Chemical Week*, vol. 161, August 11, 1999, p. 41. Sissell reports that global carbon dioxide emissions declined by 0.5 percent in 1998 despite 2.5 percent worldwide economic growth. According to the environmental group Worldwatch Institute, this proves that nations can reduce carbon emissions without jeopardizing economic growth.

Spencer, Peter. "Warmer Winters." *Consumers' Research Magazine*, vol. 83, December 2000, p. 43. Spencer discusses a report by the Cato Institute pointing to the benefits of global warming, which may include shorter winters, fewer severe winter storms, and reduced temperature variability.

Stapleton, Stephanie. "A Hot Future." *American Medical News*, vol. 43, November 6, 2000, pp. 36–37. Stapleton examines new challenges that physicians may face as a result of global warming, such as treatment of unexpected outbreaks of mutant strains of malaria, encephalitis viruses, and Lyme disease.

Staropoli, John F. "The Public Health Implications of Global Warming." *The Journal of the American Medical Association*, vol. 287, May 1, 2002, p. 2,282. This introductory article calls on health care professionals, in the absence of U.S. leadership on global warming, to become advocates of action to combat global warming. Because global climate change threatens to unleash a host of problems that may adversely affect human health, such advocacy fits in well with the growing ethos of preventive medicine within the health care community.

Stoffer, Harry. "Makers Plot CO_2 Counterattack." *Automotive News*, vol. 76, July 15, 2002, p. 137. Automakers are considering a challenge to a California law requiring the introduction of cars that meet strict emissions standards. The car manufacturers may ask voters to overturn the legislation in a referendum.

Stokes, Bruce. "A Chance to Rewrite History." *National Journal*, vol. 33, November 3, 2001, pp. 3,426–3,428. The percentage of Americans who thought that dealing with global warming was a priority dropped from 44 to 31 following the September 11, 2001 terrorist attacks on the United States. Despite this, Stokes sees the Bush administration's attempts to form an international coalition to fight terrorism as a new opportunity for the nation to adopt a multilateral response to other international problems, including environmental issues.

Stoss, Frederick W. "The Heat Is On! U.S. Global Climate Change Research and Policy." *Econtent*, vol. 23, August/September 2000, pp. 36–44. Stoss profiles efforts by the IPCC to determine why the climate is changing and what can be done about it.

Sudetic, Chuck. "As the World Burns." *Rolling Stone*, September 2, 1999, pp. 97–106 ff. In this wide-ranging article, Sudetic discusses the state of climate change research, the impact that global warming is having on the Earth and its people, and the refusal of the U.S. government to take meaningful steps to confront the problem.

———. "The Good News About Global Warming." *Rolling Stone*, October 12, 2000, pp. 80–82. This article examines James Hansen's controversial paper in which he suggested more attention be given to curbing emissions of greenhouse gases other than carbon dioxide.

Taverna, Michael A. "U.S. Seeks to Regain Edge in Climate Issue." *Aviation Week & Space Technology*, vol. 157, July 8, 2002, pp. 62–63. Taverna reports on the U.S. government's funding of new space-, air-, and surface-based based hardware designed to observe climate systems, despite the Bush administration's failure to endorse the Kyoto Protocol on climate change.

Thornton, Emily, et al. "Toyota's Green Machine." *Business Week*, December 15, 1997, pp. 108ff. The authors report on Toyota's Prius, the first mass-produced hybrid car powered by a combination of gasoline and electricity.

Toyota hopes that sales of the car will be driven by rising gas prices and by tough emissions laws designed to fight pollution and global warming.

"Too Few Fish in the Sea." *Time*, vol. 143, April 4, 1994, p. 70. This article explains the depletion of fish populations in some areas caused by gross overfishing by the commercial fishing industry. Among the other dangers to fish populations is an experiment that uses undersea sonic waves to measure global warming.

Trenberth, Kevin E. "Stronger Evidence of Human Influences on Climate: The 2001 IPCC Assessment." *Environment*, vol. 43, May 2001, pp. 8–10ff. Trenberth, head of the Climate Analysis Section of the National Center for Atmospheric Research in Colorado, provides an overview of the IPCC process and a summary of the main findings of the Third Assessment from the IPCC's Working Group I, published in January 2001. The report reaffirmed previous findings that the Earth is warming, and it is most likely caused by human activity. The lengthy article includes sidebars and footnotes.

Tutangata, Tomari'i. "Sinking Islands, Vanishing Worlds." *Earth Island Journal*, vol. 15, Summer 2000, p. 44. Also available online (URL: http://www.earthisland.org/eijournal/sum2000/ov_sum2000sinkingisles.html). Tutangata, director of the South Pacific Regional Environment Program, reports on the effects of global warming on the Pacific Ocean's 22 island countries and territories, which have contributed only 0.06 percent to total global greenhouse emissions but face complete inundation from rising sea levels.

Van Putten, Mark. "Climate Change Hits Home." *National Wildlife*, vol. 40, February/March 2002, p. 7. Van Putten discusses the ways in which global warming has had an increasing impact on people and their everyday lives, resulting in heightened awareness of the problem of climate change. Among the global warming–induced induced problems affecting the United States are severe droughts that devastate rangeland and parch forests in the West, melting permafrost in Alaska, and the increasing range of invasive animal species such as fire ants.

Veizer, Jan. "Global Warming: Armageddon or Bust?" *Forum for Applied Research and Public Policy*, vol. 16, Summer 2002, pp. 57–63. Veizer discusses the carbon cycle and the effects of increased carbon dioxide emissions on it. He also examines the evidence for global warming and questions how much of this is natural and how much is caused by humans. The truth, he concludes, lies somewhere between the views of the skeptics and the views of those who believe global warming will lead to worldwide disaster.

Walter, Katie. "The Outlook Is for Warming, with Measurable Local Effects." *Science and Technology Review*, July/August 2002, pp. 4–12. Also available online (URL: http://www.llnl.gov/str/JulAug02/Santer.html). The author discusses the content of the IPCC Third Assessment Report and cites evidence that the effects of global warming can now be directly observed on the regional level.

Wilson, Amy. "The Latest Cool Thing: Carbon Dioxide." *Automotive News*, vol. 76, October 8, 2001, p. 81. When the use of chlorofluorocarbons (CFCs) was drastically reduced by the Montreal Protocol in 1987, auto manufacturers switched to the use of Freon in car air conditioning systems. Although less environmentally destructive than CFCs, the release of Freon could still cause damage to the atmosphere. Now automakers claim they have developed the technology to compress carbon dioxide without putting additional amounts of the greenhouse gas into the atmosphere. Backers claim that the new technology even improves the performance of air conditioning systems and can boost fuel economy.

"Withering Heats." *Natural History*, vol. 101, September 1992, pp. 2ff. This article discusses ominous signs that the Earth is entering a period of global warming brought on by human activities, which may increase the extinction rates of many threatened species of animals.

Wohlforth, Charles. "Forest Killers." *Alaska*, vol. 68, March 2002, pp. 32–35. Scientist Ed Berg, who works at Alaska's Kenai National Wildlife Refuge, believes that a plague of spruce bark beetles that decimated 4 million acres of forestland in south-central Alaska in the 1990s was a direct result of global warming. According to him, a string of unusually warm summers allowed the beetles to mature in one year instead of two, doubling their assault on the forest the following spring. Experts worry that the changing climate will allow the beetle to expand its range and attack trees in areas where it has never been documented before, or it will prompt increased outbreaks of other insects.

Zimmer, Carl. "Carbon Cuts and Techno-Fixes." *Discover*, vol. 19, June 1998, pp. 60–71. The author posits several ways to stop human-induced global warming. These include using less water and energy at home, driving more fuel-efficient vehicles, recycling industrial waste, and switching to solar power.

INTERNET DOCUMENTS

Bernstein, Lenny. "Climate Change and Ecosystems." George C. Marshall Institute. Available online. URL: http://www.marshall.org/pdf/materials/89.pdf. Posted on August 2002. This essay recognizes two basic categories of ecosystems: intensively managed (farmland and managed forests and grasslands) and lightly impacted (unmanaged, natural wildlife areas). Bernstein reports that intensively managed areas will benefit from warmer temperatures and the fertilizing effect of carbon dioxide, even in developing nations. Concerning lightly impacted ecosystems, he writes that ongoing habitat destruction is more harmful to wildlife than climate change and that there is not enough evidence to support claims that global warming will lead to widespread extinction of plant and animal species.

VIDEOS

"Building the Brookhaven House: D.O.E.'s Prototype Passive Solar Home." Directed by Kirk Simon. Oley, Pa.: Bullfrog Films, 1981. Video (VHS). 25 minutes. Using live footage, graphics, and animated blueprints and diagrams, this video documents the design and construction of the Department of Energy's prototype energy efficient, passive solar, single-family home. It demonstrates the viability of using standard building materials and techniques to construct a home that is not reliant on fossil fuels and that can reduce fuel bills by 75 percent.

"Cooperating for Clean Air: Sweden's All-Out Assault on Acid Rain and Global Warming." Directed by Janine Selendy. Oley, Pa.: Bullfrog Films, 1993. Video (VHS). 38 minutes. This video documents the work of Swedish biology teacher Christer Agren, whose campaign against acid precipitation led to a 70 percent reduction of Sweden's sulfur dioxide emissions.

"The Energy Conspiracy." New York: Filmmakers Library, 1999. Video (VHS). 60 minutes. This video approaches the problem of global warming from the perspective that the fossil fuel industry has used underhanded tactics to suppress support for the development of renewable energy and boost the use of coal, oil, and gas. Among these strategies is the Leipzig Declaration, a major declaration often cited by global warming skeptics as proof that no scientific consensus exists on the subject of climate change. Investigative journalists, however, found that the majority of signatories were not climate specialists but dentists, insectologists, lab assistants, civil engineers, and weather forecasters.

"Global Warming." Wynnewood, Pa.: Schlessinger Media, 1993. Video (VHS). 30 minutes. Part of the Earth at Risk Environmental Video Series, this documentary looks into international efforts to find a solution to the problem of global warming.

"Global Warming: Hot Enough for You?" Madison, Wis.: Knowledge Unlimited, 2000. Video (VHS). 15 minutes. This video uses live footage, diagrams, graphics, and commentary from teenagers to introduce the causes and potential effects of global warming. It comes with a guidebook to help science teachers integrate the video into their classwork.

"Global Warming: Turning Up the Heat." Directed by Daniel Zuckerbrot. Oley, Pa.: Bullfrog Films, 1996. Video (VHS). 46 minutes. This program provides an overview of global warming and its effects on the Earth's climate system. Attention is given to pledges made by world leaders at the

1992 Earth Summit in Brazil to reduce emissions to 1990 levels by 2000. The documentary poses the question why, as of 1996, little had been done to carry out this promise.

"Once and Future Planet." Oley, Pa.: Bullfrog Films, 1990. Video (VHS). 24 minutes. This video follows a joint U.S./Soviet research venture in which climate scientists study the role the oceans play in the carbon cycle, the production of clouds, the influence of volcanic dust and human-made pollutants on cloud formation, photochemistry, and the production of methane. A portion of the documentary is dedicated to concrete actions people can take to improve the atmosphere by reducing reliance on fossil fuels, such as using energy more efficiently and developing renewable energy technologies.

"Rising Waters: Global Warming and the Fate of the Pacific Islands." Directed by Andrea Torrice. Oley, Pa.: Bullfrog Films, 2000. Video (VHS). 57 minutes. This video explores the effects of rising sea levels on Pacific island nations, exploring the physical and cultural impacts that have already occurred and providing a look at future problems, including the possibility that rising waters could affect New York City, the Florida Keys, and other regions in the United States. The documentary includes the personal stories of Pacific Islanders, interviews with scientists and coverage of world conferences where the problem of sea level rise has been debated.

"Silent Sentinels." Directed by Richard Smith. Oley, Pa.: Bullfrog Films, 1999. Video (VHS). 57 minutes. Smith poses questions about whether the unprecedented bleaching of tropical coral in 1998 provides evidence that the Earth's temperature is rising. He also explores the effect of rising acidity in the world's oceans on coral organisms. This wide-ranging documentary includes footage from three oceans, Australia's Great Barrier Reef, the Maldives, the Red Sea, the United Sates, and the Caribbean.

"Turning Down the Heat: The New Energy Revolution." Directed by Jim Hamm. Oley, Pa.: Bullfrog Films, 1999. Video (VHS). 46 minutes. This video discusses energy conservation and the development of alternative, renewable energy technologies as a means to fight global warming. Among the solutions profiled are solar energy in the Netherlands, Japan, and California; biogas (a naturally occurring mixture of carbon dioxide and methane) in Denmark and Vietnam; wind energy in the Netherlands and India; and hydrogen fuel cells and ground source heat in Canada.

ECONOMIC ISSUES

BOOKS

Barker, Terry, Paul Ekins, and Nick Johnstone, eds. *Global Warming and Energy Demand.* Global Environmental Change Series. New York: Routledge,

1995. This book collects revised papers from a September 1995 workshop, as well as new contributions. The essays consider a wide range of views on the use of economic measures to control greenhouse gas emissions, including energy taxes and price changes designed to abate emissions.

Cline, William R. *The Economics of Global Warming.* Washington, D.C.: Institute for International Economics, 1992. Most analyses of the economic aspects of climate change focus on the warming that may occur over the next century or so. Cline, recognizing that the process will continue for at least the next two or three centuries, extends his analysis of the damage that may be caused by global warming to cover the next 300 years, producing damage figures that far exceed standard estimates. The book concludes with a cost-benefit estimate for international action and suggestions for policy measures that can be taken to combat global warming.

Moore, Thomas Gale. *Climate of Fear: Why We Shouldn't Worry About Global Warming.* Washington, D.C.: Cato Institute, 1998. Moore, an economist, suggests that the best way to deal with global warming is to maintain the status quo, continue research on climate, and help poor countries improve their economies. Although not a climatologist, he contends that if the planet is warming at all, it will not be to the extent that climate scientists believe, and any small temperature rise will provide the benefits of lower heating bills, longer agricultural growing seasons, and improved health. Attempts to reduce fossil fuel emissions, Moore maintains, will be extremely expensive and will have adverse consequences on the U.S. economy. Focusing almost entirely on the United States, this book fails to consider the effects of sea level rise on low-lying developing countries.

Nordhaus, William D., ed. *Economics and Policy Issues in Climate Change.* Washington, D.C.: Resources for the Future, 1998. Resources for the Future is a nonprofit organization established in 1952 to conduct research into and educate the public about issues concerning natural resources and the environment. The essays in this book, the result of a 1996 workshop held in Colorado, analyze the potential economic factors involved in climate change, including the costs of greenhouse gas abatement, the costs of carbon dioxide emissions reductions, and the costs of the damage that global warming would cause if no abatement strategies were implemented. Each essay is followed by endnotes, a list of references, and commentary written by experts in the particular field covered.

Nordhaus, William D., and Joseph Boyer. *Warming the World: Economic Models of Global Warming.* Cambridge, Mass.: MIT Press, 2000. The authors present a pair of economic data models designed to help analysts and decision makers understand likely future outcomes of a variety of plans to deal with global warming. The models, called RICE-99 (Regional Dynamic Integrated Model of Climate and the Economy) and

DICE-99 (Dynamic Integrated Model of Climate and the Economy) build on earlier work by the authors, who hope these tools can help policy makers design better economic and environmental policies.

Rao, P. K. *The Economics of Global Climate Change.* Armonk, N.Y.: M. E. Sharpe, 2000. Rao writes from the premise that environmental and economic security are inextricably linked; ignoring either issue, particularly when it comes to global warming, may lead to disaster. Complicating matters is that economics and climate studies are both inexact sciences rife with uncertainties, risk, and incomplete information. Rao tries to make sense of this difficult situation by exploring the scientific background of climate change, analyzing how decisions can be made in the midst of such uncertainty, and suggesting integrated solutions that may lead to win-win situations for both the environment and the economy. Researchers should be forewarned that this is highly technical material recommended only for those with a specific interest in economic studies. However, general readers will be able to use the glossary, the list of web sites, and the extensive excerpts from the United Nations Framework Convention on Climate Change and the Kyoto Protocol.

ARTICLES

Avis, Ed. "Green Advocates See Dollar Signs." *Crain's Chicago Business*, vol. 23, September 18, 2000, p. SR5. Avis reports on the Midwest Global Warming Leadership Council. Coordinated by the Environmental Law and Policy Center in Chicago, the council is a group of companies that believe environmentally sound development will help the Midwest economy.

"The Climbing Cost of Climate Change." *Earth Island Journal*, vol. 16, Summer 2001, p. 19. This article discusses the ramifications of a report issued by the United Nations Environmental Programme's financial services initiative predicting that damages caused by global warming could cost the world as much as $304.2 billion per year unless efforts are made to curb emissions of greenhouse gases. According to the report, low-lying nations such as the Maldives and the Marshall Islands may, by the year 2050, be forced to dedicate more than 10 percent of their gross domestic product (GDP) to fighting the effects of rising sea levels. Worldwide, the cost to defend homes, factories, and power stations from rising sea levels and storm surges may reach $47 billion by 2050, while damages from ecosystem losses could be as high as $70 billion. Damage brought about by droughts, floods, and fires, as well as by more frequent cyclones and hurricanes, could adversely affect agriculture and forestry. Other areas of concern include losses to the construction, transportation, and tourism industries, and higher levels of mortality and health costs.

Dhar, Arunima. "Insurance Companies Warn Global Warming Will Cost $70 Billion Annually." *World Watch*, vol. 16, January/February 2003, p. 11. During the UN talks on climate change in New Delhi, India, in October 2002, representatives from Germany-based Munich Re (one of the world's largest insurance companies) reported that as of September 2002, natural disasters had cost countries and communities an estimated $56 billion during that year. Company experts attributed the vast majority of the damages to climate change.

Goldsmith, Edward, and Caspar Henderson. "The Economic Costs of Climate Change." *The Ecologist*, vol. 29, March/April 1999, pp. 98–100. The authors contend that the long-term costs of doing nothing about global warming will far outweigh immediate mitigation costs that many industrialists are strenuously trying to avoid.

"Kyoto Protocol Could Cripple Industry." *Chemical Market Reporter*, vol. 253, June 15, 1998, p. 9. This article reports on an analysis by the economic consulting firm Charles River Associates estimating that the Kyoto Protocol could be 10 times more costly to U.S. consumers than the Clinton administration claimed.

Levy, David L. "Business and Climate Change: Privatizing Environmental Regulation?" *Dollars & Sense*, January/February 2001, pp. 21–23ff. Levy, a management professor at the University of Massachusetts in Boston, examines the arguments against international regulations on the emission of greenhouse gases.

"An Optimal Transition Path for Controlling Greenhouse Gases." *Science*, vol. 258, November 20, 1992, pp. 1,315ff. This article presents a Dynamic Integrated Climate Economy (DICE) model for controlling greenhouse gases, suggesting that a modest carbon tax would be an efficient approach to slow global warming.

Ottman, Jacquelyn A. "Green Marketing: A 'Cool' Strategy to Stop Global Warming and Corporate Abuses." *In Business*, vol. 24, July/August 2002, pp. 30–32. In the absence of real leadership from the U.S. government on the issue of global warming, Ottman examines the possibility that advertisers and marketers could make it "cool" for Americans to take a stand on climate change by using the same communications techniques that made it cool to stop smoking, wear seat belts, and stop littering. She presents a list of incentives for advertisers to do so, as well as a list of four marketing techniques.

Pistorio, Pasquale, and Mary Hager. "Global Warming: Business' Biggest Economic Threat." *Chief Executive*, October 2001, p. 18. Pistorio, president and chief executive officer (CEO) of a semiconductor manufacturing corporation, urges CEOs of other companies to consider the economic benefits of sustainable business practices.

Schelling, Thomas C. "The Cost of Combating Global Warming: Facing the Tradeoffs." *Foreign Affairs*, vol. 76, November/December 1997, pp. 8–14. Schelling argues that the costs of mitigating climate change will be borne primarily by western Europe, North America, and Japan, but the beneficiaries will be future generations in the developing world. Because of this, any international action to combat global warming will amount to a foreign aid program.

Scott, Alex. "BP Experiments with CO_2 Emissions Trading." *Chemical Week*, vol. 160, October 28, 1998, p. 42. Scott reports on steps British Petroleum took to institute its own emissions trading system even before parties to the Kyoto Protocol met to finalize international and national systems.

"Truth and Fiction About Global Warming." *Iron Age New Steel*, vol. 14, July 1998, p. 84. This article reports on worries within the steel industry that acceptance of the Kyoto Protocol by the United States would cause steel production to shift from North America to developing countries that would not be forced by the treaty to reduce their emissions.

Weidenbaum, Murray. "Curbing Global Warming: Is the Cost Too High?" *USA Today*, vol. 127, July 1998, pp. 68–70. With an emphasis on the uncertainties about climate change prediction and global warming theory, Weidenbaum discusses various strategies that have been proposed to reduce carbon dioxide emissions.

Wysham, Daphne. "The World Bank: Funding Climate Chaos." *The Ecologist*, vol. 29, March/April 1999, pp. 108–110. Wysham reports on how the World Bank has spent billions of dollars building fossil fuel–related projects in developing countries, which will radically increase the total amount of carbon dioxide emitted around the world. In the meantime, the bank plans on profiting from these new emissions by entering the market in emissions trading.

Internet Documents

Coalition for Environmentally Responsible Economies. "Values at Risk: Climate Change and the Future of Governance." Available online. URL: http://www.ceres.org/pdf/climate.pdf. April 2002. This report examines the potentially devastating economic consequences of climate change and provides advice for corporate directors seeking responsible approaches to their duties in the face of threats posed by climate change.

Partnership for Climate Action. "Common Elements Among Advanced Greenhouse Gas Management Programs." Environmental Defense Fund. Available online. URL: http://www.environmentaldefense.org/documents/1885_PCAbooklet.pdf. April 2, 2002. The Partnership for Climate Action provides a forum for corporations to explore viable ways to confront the problem of climate change. Member companies include Alcan, BP, DuPont, Ontario Power Generation, Shell International, and Suncor Energy. This report shows how these companies have incorporated greenhouse gas reduction strategies into their business practices and sheds light on what they have learned in the process.

World Bank. "Poverty and Climate Change: Reducing the Vulnerability of the Poor." Available online. URL: http://lnweb18.worldbank.org/ESSD/essdext.nsf/46DocByUnid/6449D122940C7A9485256C4F005349D7/$FILE/PovertyAndClimateChange2002.pdf. October 2002. This report, presented at the Eighth Conference of Parties to the UNFCCC (COP 8), in New Delhi, India, seeks solutions to the question of how to integrate adaptation to climate change into poverty reduction.

ETHICS

BOOKS

Brown, Donald A. *American Heat: Ethical Problems with the United States' Response to Global Warming.* Studies in Social, Political, and Legal Philosophy series. Lanham, Md.: Rowman and Littlefield, 2002. This book examines the ethics of the positions the United States has taken in global warming negotiations. Brown concludes that by any standards, U.S. policy in this area is ethically bankrupt.

Coward, Harold, and Thomas Hurka, eds. *Ethics and Climate Change: The Greenhouse Effect.* Waterloo, Ontario, Canada: Wilfrid Laurier University Press, 1993. In a series of essays, scientists, social scientists, humanists, legal and environmental scholars, and corporate researchers offer ethical analysis of possible responses to the problem of global warming. Contributors assess personal, corporate, government, and international responsibility for climate change, and recommend ways that decision makers can determine solutions and apportion responsibility.

ARTICLES

"Ethical Concerns for the Global-Warming Debate." *The Christian Century,* vol. 109, August 26, 1992, pp. 773ff. This article argues that global warming is not only a scientific debate but also a moral issue. It outlines six key concerns that must be brought to the ethics debate on global warming: distinguishing between global warming and global climate change; avoiding moral compartmentalization and seeking a just integration of proposed solutions to the ecological and social crises; advocating frugality; recognizing that international cooperation is an ecological imperative; giving fair consideration to all parties with stakes in the outcome; and interpreting the problem in the context of the morality of high risk taking.

Gardner, Christine J. "U.S. Churches Join Global Warming Debate." *Christianity Today*, vol. 42, October 5, 1998, p. 16. In October 1998, mainline Protestant and evangelical leaders gathered in Columbus, Ohio, for the Midwest Interfaith Climate Change conference to prepare a grassroots lobbying effort in support of the Kyoto Protocol. Attendees expressed their belief that environmental stewardship is an act of compassion toward the poor.

INTERNET DOCUMENTS

Union of Concerned Scientists. "Common Sense on Climate Change: Practical Solutions to Global Warming. Available online. URL: http://www.ucsusa.org/global_environmental/global_warming/page.cfm?pageID=793. Updated on February 6, 2003. 2002. The Union of Concerned Scientists (UCS) urges policy makers to take an ethical approach to the use of science and technology. This essay urges businesses, policy makers, and consumers to take what the UCS calls commonsense steps to fight global warming, including improving fuel efficiency in cars and SUVs, modernizing the U.S. electricity system, and protecting threatened forests.

POLICY AND POLITICS

BOOKS

Chambers, Bradnee W. *Inter-Linkages: The Kyoto Protocol and International Trade and Investment Regimes.* Tokyo: United Nations University Press, 2001. The nine essays and discussion papers in this book focus on the links between the Climate Change Convention on one hand and the World Trade Organization, international investment agreements, and private and contractual trade law on the other. As a whole, the book addresses the need for a greater awareness of potential impact of climate change and for a broader view on the context of international laws and policies. Among the topics examined are the potential role of the private sector and issues related to noncompliance.

Kelly, Robert C. *The Carbon Conundrum: Global Warming and Energy Policy in the Third Millennium.* Houston, Tex.: CountryWatch, 2002. Kelly is founder of CountryWatch, a company that provides country-specific geopolitical information for the corporate, government, and academic markets. In this book, he presents a comprehensive overview of the global warming controversy. Topics covered include a history of scientific debate surrounding climate change; evidence of historic changes in the Earth's climate; an assessment of three alternative scenarios for the interaction between global climate and global economy in the 21st century and beyond; a review of future energy supply options; and suggestions for implementing what Kelly believes is the optimal solution to the problem.

Kursunoglu, Behram N., Stephan L. Mintz, and Arno Perlmutter, eds. *Global Warming Energy and Policy.* New York: Plenum, 2001. This book represents proceedings of a symposium on global warming and energy policy. In exploring the theory that the globe will become warmer as a result of carbon dioxide emissions, contributors outline problems associated with predicting temperature rise, examine the effect of rising temperature on economic activity, discuss steps that can be taken to clarify the global warming debate, and discuss what society might do while waiting for these steps to produce results. Other essays look into the scientific uncertainties of determining the health effects of the fine particles emitted as a by-product of fossil fuel combustion.

Minger, Terrell J., ed. *Greenhouse Glasnost: The Crisis of Global Warming.* New York: Ecco Press, 1990. The essays in this book are adapted from talks given at Greenhouse/Glasnost: The Sundance Symposium on Global Climate Change, organized by actor Robert Redford's Institute for Resource Management to bring together U.S. and Soviet scientists, policy makers, environmentalists, industry leaders, and artists to discuss global warming. Contributors include astronomer Carl Sagan, atmospheric physicist Georgii S. Golitsyn, biologist James E. Lovelock, and U.S. senator Bill Bradley. Each essay is preceded by a biographical essay about the author, and many are followed by a list of references. The book opens with an introduction by Redford and closes with photographs from the symposium and an open letter to Mikhail Gorbachev and George H. W. Bush urging them to take action on global warming.

Oberthur, Sebastian, et al. *The Kyoto Protocol: International Climate Policy for the 21st Century.* New York: Springer Verlag, 2000. This book offers an inside perspective on the debate about the Kyoto Protocol, including a scholarly analysis of the history and content of the protocol itself as well as of the economic, political, and legal implications of its implementation.

Paterson, Matthew. *Global Warming and Global Politics.* Environmental Politics Series. New York: Routledge, 1996. Paterson examines major theories within the discipline of international relations and considers how they might be able to provide accounts of the emergence of global warming as a political issue. The book covers negotiations leading up to the 1992 United Nations Framework Convention on Climate Change in Rio.

Rosenberg, Norman J., et al., eds. *Greenhouse Warming: Abatement and Adaptation.* Washington, D.C.: Resources for the Future, 1989. This book consists of technical papers presented at a 1988 global warming workshop organized by the Climate Resources Program at Resources for the Future, an independent nonprofit organization dedicated to advancing research and public education in the quality of the environment and in the sustainable use of natural resources. Among the subjects covered are the causes and impacts of the greenhouse effect, the regional consequences of and responses to sea level rise, adapting agriculture to climate change, and potential strategies for adapting to global warming. Contributors include economists, geographers, environmental scientists, and public policy experts.

Taylor, Prue. *An Ecological Approach to International Law: Responding to Challenges of Climate Change.* New York: Routledge, 1998. This book explores the question of why international law has failed to respond to the problem of global climate change, offering a major critique of international law based on environmental ethics. Taylor examines particular international legal responses to global climate change, concluding that current international environmental law is fundamentally flawed and not equipped to meet global challenges.

Victor, David G. *The Collapse of the Kyoto Protocol and the Struggle to Slow Global Warming.* Princeton, N.J.: Princeton University Press, 2001. The author takes the position that the Kyoto Protocol is fundamentally flawed because it has set unrealistic high goals for capping emissions and because it includes too many loopholes that allow nations to avoid complying with these goals. After presenting an in-depth look at the protocol's flaws, Victor (who is not against taking action to mitigate the negative effects of global warming) not only declares that the treaty's basic architecture needs to be rethought but also suggests several market-based ways to do so. Although this book focuses primarily on the mechanics of the Kyoto Protocol, it does include an appendix that briefly describes the causes and effects of global warming.

ARTICLES

Aronson, Adam L. "From 'Cooperator's Loss' to Cooperative Gain: Negotiating Greenhouse Gas Abatement." *The Yale Law Journal*, vol. 102, June 1993, pp. 2,143ff. This long, technical article discusses the problems involved in developing national and international greenhouse gas abatement strategies. Aronson suggests that an international tradable emissions permit system might be the best option, as it would promote abatement at the lowest possible cost, prove more administrable and equitable than a tax system, and minimize the welfare loss from uncertainty about the level of marginal costs.

Bailey, Scott. "TNRCC Is Weighing 'Historic' Decision on Global Warming." *San Antonio Business Journal*, vol. 15, November 23, 2001, p. 3. Bailey reports on efforts by the Texas Natural Resources Conservation Commission (TNRCC) to draft a long-term global warming plan in the midst of lobbying efforts, on one hand, by environmental groups that rallied together and submitted a petition urging the TNRCC to combat global warming and, on the other hand, by major automakers, oil and mining firms, and electrical utilities, who argue that new rules and regulations could raise the costs of and reduce the demand for their products.

Baliunas, Sallie. "Regulation Based on Science." *Vital Speeches of the Day*, vol. 68, April 15, 2002, pp. 390–395. This is a transcript of a speech given at Hillsdale College in Michigan by Baliunas, an astrophysicist at the Harvard-Smithsonian Center for Astrophysics and deputy director at southern California's Mount Wilson Observatory. In it, Baliunas claims that the Kyoto Protocol is fatally flawed on scientific, economic, and political grounds. Among the arguments presented are claims that solar and wind power are unreliable for large-scale energy generation; that the use of fossil fuels has had a positive effect on the environment and has improved the longevity, health, welfare, and productivity of humans; and that reducing emissions would strain international relations by wreaking havoc with the U.S. economy, thereby adversely affecting developing nations that rely on the United States to maintain stability and provide aid. Baliunas also questions the validity of computer models that predict catastrophic anthropogenic warming, instead suggesting that warming trends during the 20th century may have been caused by fluctuations in the Sun's energy output.

Blankinship, Steve. "Bush Says U.N. Global Warming Summary Slants Report's Findings." *Power Engineering*, vol. 105, November 2001, p. 67. Blankinship reports on the Bush administration's claim that United Nations Intergovernmental Panel on Climate Change's 2001 Synthesis Report "does not sufficiently acknowledge the uncertainties involved in climate change science and assessment" that are reflected in the Third Assessment Report as a whole. President Bush contends that Robert Watson, the IPCC's chairman at the time, intentionally slanted the summary report to support the theory of human-induced climate change.

Bush, George W. "Global Climate Change: Making Commitments We Can Keep and Keeping Commitments We Can Make." *Vital Speeches of the Day*, vol. 67, July 1, 2001, pp. 546–548. This is the complete text of the speech given by President Bush at the White House on June 11, 2001, in which he refers to the Kyoto Protocol as "fatally flawed in fundamental ways." He goes on to commit the United States to "work within the U.N. framework and elsewhere to develop with our friends and allies and nations throughout the world an effective and science-based response to the issue of global warming."

"Bush Unveils Voluntary Response." *Occupational Hazards*, vol. 64, April 2002, pp. 17–18. This article covers George W. Bush's speech in which he presented his business-friendly, voluntary alternative to the Kyoto Protocol. An administration spokesman admitted, however, that the plan would cause greenhouse gas emissions in the United States to rise even if industry meets the voluntary targets.

"Climate Change Belly-Flop." *Oil and Gas Journal*, vol. 99, June 18, 2001, p. 19. This article takes a critical look at the National Academy of Sciences report on global warming commissioned by the Bush administration. A major point of contention is the report's opening sentence, which asserts that "greenhouse gases are accumulating in Earth's atmosphere as a result of human activities, causing surface air temperatures and subsurface ocean temperatures to rise." This statement is not qualified until the third sentence ("The changes observed over the last several decades are likely mostly due to human activities, but we cannot rule out that some significant part of these changes are also a reflection of natural variability"), which the author of this article complains has largely been ignored by the media, as has the fact that the rest of the report elaborates that uncertainty.

"Climate Scientists Advise White House on Global Warming." *Journal of Environmental Health*, vol. 64, September 2001, pp. 46–47. This article reports on the contents of "Climate Change Science: An Analysis of Some Key Questions," a report written by a committee of the National Research Council at the request of the Bush administration to provide a summary of the current understanding scientists have of global climate change. The committee concluded that the current thinking of the scientific community is accurately reflected in the IPCC's belief that the global warming that occurred during the second half of the 20th century is likely the result of increases in greenhouse gases. The report cautioned, however, that this conclusion has not been proven beyond doubt. The committee also stated its belief, contrary to strong accusations made by global warming skeptics and oil company representatives, that the IPCC's abridged technical and policy-maker summaries accurately summarize the contents of the full IPCC Working Group I report issued in 2001.

Clinton, William J. "Joint Statement on Cooperation to Combat Global Warming." *Weekly Compilation of Presidential Documents*, vol. 36, June 12, 2000, pp. 1,277–1,278. This is the text of a joint statement between U.S. president Bill Clinton and Russian federation president Vladimir Putin reaffirming the commitment of both nations to cooperate in taking multilateral action to reduce the serious risks of global warming. Clinton and

Putin both declare their conviction that national and global economic growth can be achieved while steps are taken to stop global warming.

———. "Remarks During the White House Conference on Climate Change." *Weekly Compilation of Presidential Documents*, vol. 33, October 13, 1997, pp. 1,496–1,501. In this transcript of a panel discussion on the science of global warming and climate change, President Clinton, Thomas Karl (senior scientist at the National Oceanic and Atmospheric Administration's National Climatic Data Center), Diana Liverman (chair of the National Academy of Sciences Committee on Human Dimensions of Climate Change), and others talk about evidence that the temperature of the Earth is rising and that this increase is affecting people's lives.

"Clinton Unveils New 'Greenhouse' Policy." *Science News*, vol. 144, October 23, 1993, p. 263. This article outlines President Clinton's Change Action Plan for averting the threat of global warming. The plan involves about 50 voluntary measures to reduce greenhouse gas emissions with the goal of returning net U.S. emissions to 1990 levels as stipulated by the 1992 United Nations Framework Convention on Climate Change. Measures include creating new efficiency standards for household appliances, promoting use of energy-efficient lighting systems in offices, and placing tighter regulatory controls on the release of methane from landfills. Environmentalists applauded the plan but expressed disappointment that the measures were not mandatory.

Crook, Clive. "Bush Broke His Promise on Global Warming." *National Journal*, vol. 34, March 2, 2002, pp. 595–596. President Bush, who has been quoted as saying he believes global warming is a problem that must be dealt with, rejected the Kyoto Protocol with the promise that he would provide an alternative approach. In this article, Crook, who agreed that the Kyoto Protocol should be scrapped, refers to Bush's plan as a "joke" that, while pretending otherwise, will lead to increased emissions and that does nothing to face the long-term problem of global warming. Crook offers his own alternatives to both plans, including instituting either a carbon tax or a system of tradable emissions permits.

DeMuth, Christopher. "The Kyoto Treaty Deserved to Die." *The American Enterprise*, vol. 12, September 2001, p. 6. DeMuth prematurely predicts the death of the Kyoto Protocol, which remained viable on an international level despite its rejection by the Bush administration. As an alternative to Kyoto, DeMuth offers a three-pronged approach to the problem of global warming: think globally by conducting serious, depoliticized scientific research on the global aspects of climate change; act locally by rejecting international treaties and implementing local, regional, and incremental government actions; and place a temporary moratorium on regulatory controls that may damage the economy.

Annotated Bibliography

Dickey, Christopher. "The Politics of Apocalypse." *Newsweek*, July 23, 2001, p. 49. According to Dickey, because the threat of global warming can be blamed on multinational corporations that produce and market carbon dioxide–producing fossil fuels, the issue of anthropogenic climate change has added a sense of coherence to the growing protest movement against globalization. Many of these activists credit George W. Bush's anticonservation agenda with raising awareness of the importance of the global warming problem.

Dickey, Christopher, and Adam Rogers. "Smoke and Mirrors." *Newsweek*, February 25, 2002, pp. 30ff. The authors provide a critical look at George W. Bush's plan for voluntary measures to cut "greenhouse-gas intensity," defined as a ratio between emissions and gross domestic product. Environmentalists who studied the administration's own projections, however, showed that the plan would see an increase of greenhouse gas emissions of almost 30 percent, about the same amount that would occur with no plan at all.

"Faith-Based Reasoning." *Scientific American*, vol. 284, June 2001, p. 8. This editorial contends that President George W. Bush's decisions to abandon the Kyoto Protocol, reduce emission control laws, and continue the development of a national strategic missile defense system—all of which stand in direct opposition to the views of the scientific community—could result in disaster for the United States.

Gardiner, David, and Lisa Jacobson. "Will Voluntary Programs Be Sufficient to Reduce U.S. Greenhouse Gas Emissions?" *Environment*, vol. 44, October 2002, pp. 24–33. President George W. Bush, once a global warming skeptic, has acknowledged the importance of taking action to address the impacts of global climate change. On February 14, 2002, he released his Global Climate Change Initiative, which established a goal of reducing greenhouse gas emissions intensity by 18 percent by the year 2012. The proposal relies on voluntary efforts by corporations to meet this emissions target but does not attribute responsibility to specific sources of emissions. The authors analyze the proposal, concluding that it will do little to reduce greenhouse gas emissions in the United States because the reduction targets are not aggressive enough and because plans based on voluntary compliance lack necessary accountability.

Gelbspan, Ross. "Beyond Kyoto Lite." *The American Prospect*, vol. 13, February 25, 2002, pp. 26–27. Gelbspan, author of the book *The Heat Is On*, reports on the widespread impression that George W. Bush's rejection of the Kyoto Protocol even while it was signed by 165 other nations may signal a shift in the global balance of political power away from the United States and toward the European Union. While the protocol is far from perfect, according to Gelbspan, at the very least it provides an international framework for mitigating the effects of global warming.

———. "Climate Change Demands Radical Action." *Earth Island Journal*, vol. 16, Summer 2001, p. 32. Gelbspan discusses attempts by the United States to undermine the Kyoto Protocol by insisting on planting trees and buying cheap emissions credits from developing countries rather than by enacting domestic reductions in oil and coal burning. He suggests that future international negotiations focus on subsidy switches, progressive fossil-fuel efficiency standards, and a technology-transfer fund.

———. "A Modest Proposal to Stop Global Warming." *Sierra*, vol. 86, May/June 2001, pp. 62–67. Gelbspan accuses the United States of acting like a rogue state at climate talks in The Hague, the Netherlands, where many industrial nations said they were willing to cut fossil fuel emissions while the United States would commit only to planting trees and buying cheap pollution allowances from developing countries. Subsequently, the talks collapsed. Gelbspan believes that the solution to global warming is to bring the "Carbon Age" to an end by committing to a rapid switch from fossil fuels to renewable energy.

Glassman, James. "An American Alternative to Enviro-Gloom." *The American Enterprise*, vol. 12, December 2001, p. 15. After the Bush administration officially rejected the Kyoto Protocol in March 2001, the international community put enormous pressure on the United States to do something about global warming. Glassman suggests that Bush should work to switch the focus of international treaties away from reducing carbon dioxide emissions and toward ending poverty. Such an approach, which Glassman suggests could be expanded to become a central pillar of U.S. foreign policy, would give developing countries the tools they need to respond to environmental calamity and would simultaneously help the war on terrorism by promoting democratic ideals around the world.

Goldstein, Andrew. "For Bush, It's Not Easy Being Green." *Time*, vol. 159, February 25, 2002, p. 19. This article reports on the angry response among environmentalists to George W. Bush's "Clean Skies" initiative, which they believe would undermine the Clean Air Act and increase power-plant emissions, and his voluntary alternative to the rejected Kyoto Protocol, which would increase emissions of greenhouse gases by 14 percent over the next decade.

———. "Too Green for Their Own Good?" *Time*, vol. 160, August 26, 2002, pp. A58–A60. Goldstein examines the tactics of U.S. environmental groups, which are stronger and richer than ever but seem to fighting a losing battle on issues such as global warming.

Gunnell, Barbara. "Ditching the Yanks to Save Our Climate." *New Statesman*, vol. 14, April 9, 2001, p. 12. Gunnell reports on President Bush's announcement that the Kyoto Protocol is "dead" and examines the implications of this announcement for the European Union (EU) and for trade and economic relations between the EU and the United States.

Hawkins, David, and Jerry Taylor. "Should the U.S. Sign the Kyoto Protocol?" *New York Times Upfront*, vol. 134, March 25, 2002, p. 22. Hawkins, representing the Natural Resources Defense Council, and Taylor, of the Cato Institute, present opposing views about whether the United States should sign the Kyoto Protocol.

Hertsgaard, Mark. "Bush and Global Warming." *The Nation*, vol. 273, December 10, 2001, pp. 5–6. This editorial criticizes the disparity between the Bush administration's insistence that every country with a seat at the United Nations ally itself with the United States in battling terrorism and its simultaneous refusal to join the rest of the world in fighting global warming, which may, in the long run, Hertsgaard writes, pose a much greater threat to the well-being of humanity than terrorism.

Hess, Glen. "Seven States to File Clear Air Lawsuit." *Chemical Market Reporter*, vol. 263, March 3, 2003, p. 1. This article reports that attorneys general of seven states have announced their intention to file a lawsuit to force the federal government to regulate carbon dioxide emissions from power plants. The lawsuit will argue that the failure of the EPA to regulate carbon dioxide violates the Clean Air Act and contributes significantly to global warming. The chemical industry and other manufacturers oppose mandatory controls on carbon dioxide emissions, worrying that controls would force power producers to switch from coal to natural gas, which would cause increases in the cost of the key feedstock.

Kaiser, Jocelyn. "17 National Academies Endorse Kyoto." *Science*, vol. 292, May 18, 2001, pp. 1,275–1,277. Shortly after President George W. Bush rejected the Kyoto Protocol, the Royal Society of the United Kingdom organized 17 national academies of science to issue a statement affirming the IPCC's conclusion that human activities are warming the planet and urging those with doubts to ratify the protocol. The U.S. National Academy of Sciences (NAS) was invited to sign but declined on the grounds that it could not endorse a document it did not help draft.

Kerry, John. "Bush Takes a Backseat." *Time*, vol. 160, August 26, 2002, p. A49. Massachusetts senator Kerry blasts the Bush administration's rejection of the Kyoto Protocol and its lack of a viable alternative. He accuses the administration of turning "the environmental agenda over to big polluters, denouncing even modest reforms as technologically impossible and economically ruinous."

Kronick, Charlie. "The International Politics of Climate Change." *The Ecologist*, vol. 29, March/April 1999, pp. 104–107. Kronick complains that international efforts to prevent global warming are progressing too slowly and are not radical enough to stop the continued rise of greenhouse gas emissions.

Laird, Frank N. "Just Say No to Greenhouse Gas Emissions Targets." *Issues in Science and Technology*, vol. 17, Winter 2000/2001, pp. 45–52. Laird argues that the emissions targets and timetables of the Kyoto Protocol are so flawed that they actually impede efforts to deal with global warming. The United Nations and signatory nations, however, have invested so much effort in trying to make the treaty viable that they are unlikely to let it die and look for alternatives.

Late, Michele. "California Law to Tackle Auto Emissions, Global Warming." *The Nation's Health*, vol. 32, September 2002, p. 12. This article discusses the passage of a California law requiring the California Air Resources Board to develop carbon pollution standards for vehicles for model year 2009 and beyond. The legislation, which could lead to cleaner cars nationwide, was passed in response to President George W. Bush's unwillingness to take action to fight global warming. The auto industry complained that the changes will lead to fewer vehicle choices for consumers, but California governor Gray Davis predicted that the law will result in higher-quality vehicles.

Lawler, Andrew. "House Panel Icy to White House Plans." *Science*, vol. 279, February 20, 1998, p. 1,124. Lawler reports on the reluctance of Republicans in the House of Representatives to support the Clinton administration's plans to implement the Kyoto Protocol. A majority in Congress felt that the science of climate change was too vague to allocate resources to stop global warming.

Lean, Geoffrey. "Future Looks Good with 20/20 Vision." *New Statesman*, vol. 15, February 25, 2002, pp. 22–24. In February 2002, Britain's Performance and Innovation Unit (PIU) published the "Energy Review," which lays down a blueprint that would revolutionize Britain's energy policy. The report concludes that Britain could cut its carbon dioxide emissions by 60 percent by 2050 with relatively little impact on the economic growth rate. Lean believes the report could be stronger, particularly regarding the implementation of renewable energy sources, but says it represents the most fundamental change in energy policy since the Industrial Revolution. There are signs that Prime Minister Tony Blair may implement the report's plan, but the author worries that it may be weakened by fossil fuel lobbyists before it become law.

Masaki, Hisane. "COP4 to Target Third World Global Warming." *Japan Times*, vol. 38, September 7–13, 1998, p. 3. This article anticipates controversial discussions on voluntary commitments by developing countries to fight global warming at the Fourth Conference of Parties to the

United Nations Framework Convention on Climate Change (COP 4) in Buenos Aires, Argentina.

Massey, Rachel. "Global Warming: An Opportunity for World Response." *The Humanist*, vol. 61, March/April 2001, pp. 5–7. Massey argues that policy makers who try to delay action to reduce greenhouse gas emissions are not, as they claim, protecting the U.S. economy as a whole but rather the special interests of a small group of fossil fuel industries.

McKibbin, Warwick J., and Peter J. Wilcoxen. "Climate Change after Kyoto." *The Brookings Review*, vol. 20, Spring 2002, pp. 6–10. In considering ways to combat global warming, the authors attempt to find the middle ground between those who cite scientific uncertainty as a reason to take no action and those who believe that climate change should be dealt with immediately, regardless of the effects on global economy. In this article, they offer an alternative to the Kyoto Protocol that maintains the original spirit and goals of the 1992 United Nations Framework Convention on Climate Change.

Musser, George. "Climate of Uncertainty." *Scientific American*, vol. 285, October 2001, pp. 14–15. Musser discusses how the uncertainties that surround global warming and its effects on the global climate have created an unnecessary standstill among U.S. politicians who demand further proof before enacting policies that deal with the problem.

Nash, James L. "Congress Defies Bush on Arsenic, Global Warming." *Occupational Hazards*, vol. 63, September 2001, pp. 47–48. Votes cast during the 107th Congress on some key environmental issues revealed that even the Republican-controlled House was "greener" than President Bush. On July 27, the House approved an amendment stipulating that the Environmental Protection Agency must immediately adopt a high standard to reduce arsenic in drinking water that was finalized at the end of the Clinton administration and suspended by Bush. On the subject of global warming, two of the Senate's leading figures—Joseph Lieberman and John McCain—announced plans to introduce legislation that would put a cap on greenhouse gas emissions and create a market through which companies would be rewarded for creating new technologies to mitigate the problem.

Normile, Dennis. "Kickoff in Kyoto." *Popular Science*, vol. 252, March 1998, p. 26. Written only months after the United Nations Climate Change Conference in Kyoto, Japan, this article reports that the Kyoto Protocol is only the first step in long-term efforts to reduce global warming.

Paemen, Hugo. "Cautious Optimism After Kyoto." *Europe*, February 1998, pp. 28–29. Paemen analyzes criticism from both the U.S. business and environmental communities that U.S. negotiators at Kyoto, Japan, were pushovers who caved in to pressures from other countries. Paemen disagrees, claiming that U.S. negotiators "are among the toughest in the business" and that the Kyoto Protocol would not have been drafted without "strong and determined leadership from the United States."

Parry, Sam. "Hamstrung Debate on Global Warming." *Dollars & Sense*, July/August 1998, p. 19. This article reports on President Clinton's lack of action in implementing the Kyoto Protocol. Although polluting industries such as automobile manufacturers, oil and coal suppliers, and utilities have spent millions to block greenhouse gas reductions, Parry suggests that the Clinton administration should capitalize on the overwhelming public support for the treaty.

"President's Stance on Global Warming Wins Support." *Chemical Market Reporter*, vol. 261, February 25, 2002, p. 25. This anonymous article reports on industry support for and environmentalists' rejection of George W. Bush's voluntary plan to slow the increase of greenhouse emissions. The National Association of Manufacturers (NAM) called it "a measured response . . . that takes into account the need to sustain economic growth in order to invest in future environmental improvements." Environmentalists, on the other hand, say the voluntary approach "has failed for a decade now" and there is "no reason to believe polluters are suddenly going to see the light." Under this plan, they maintain, the United States will be producing 35 percent more greenhouse gases in 2010 than it would under the Kyoto Protocol.

"Promises, Promises: What Did Bush Say? What Did He Do?" *Omni*, vol. 14, June 1992, pp. 9ff. This article evaluates whether President George H. W. Bush lived up to his campaign promises concerning such environmental issues as global warming, the Clean Air Act, wetlands, and energy. The short answer to that question, according to the author, is that he did not.

Rabe, Barry G. "Statehouse and Greenhouse: The States Are Taking the Lead on Climate Change." *The Brookings Review*, vol. 20, Spring 2002, pp. 11–13. Even as the George W. Bush administration rejected the Kyoto Protocol in favor of a voluntary domestic program that many believe will cause greenhouse gas emissions to rise, several states have enacted laws establishing specific strategies to reduce emissions of carbon dioxide, methane, and other greenhouse gases. Rabe details laws passed in New Jersey, Oregon, and Massachusetts, and suggests that Bush use National Environmental Performance Partnership agreements to decentralize the fight against global warming by promoting more state-based approaches to climate change policy.

Raeburn, Paul. "Global Warming: Look Who Disagrees with Bush." *Business Week*, April 30, 2001, p. 72EU11. Raeburn reports that some of the world's biggest corporations disagree with President Bush's decision to reject the Kyoto Protocol. Among them are DuPont and Eastman Chemical, both of which see business value in being part of the solution to global warming. Meanwhile, BP Amoco, Royal Dutch/Shell, Ford, and General Motors all pulled out of the Global Climate Coalition, which had opposed mandatory limits on greenhouse emissions. BP Amoco, Royal Dutch/Shell, Alcoa, Georgia-Pacific, Toyota, and other companies have taken the additional step of joining the Pew Center on Global Climate Change, whose goal is to search for solutions to the global warming problem. The author concludes by suggesting that the probusiness Bush administration is out of step with real trends in big business.

"Realistic Mitigation Options for Global Warming." *Science*, vol. 257, July 10, 1992, pp. 148ff. This article outlines a framework for evaluating global warming mitigation options, with an emphasis on cost-effectiveness as an essential guideline in evaluating and comparing alternative mitigation strategies.

Richter, Burton. "Global Warming: What Comes After Kyoto?" Kyoto?" *The OECD Observer*, August 2002, pp. 35–37. Richter examines the potential impact of the Kyoto Protocol, concluding that meeting the goals of the treaty will have little impact on greenhouse gas emissions. Policy makers seem unwilling to take meaningful action since they rarely look beyond the next election. Any effective mitigation strategies, on the other hand, would take decades to have a real impact on emission levels. Richter warns, however, that the link between energy and the economy means that the investment necessary to bring global warming under control will become greater the longer a significant change is delayed.

Sandler, Todd. "After the Cold War, Secure the Global Commons." *Challenge*, vol. 35, July/August 1992, pp. 16ff. Sandler suggests that, with the Cold War over, nations turn their attention to dealing with some of the problems of the global commons, including ozone depletion, global warming, deforestation, and acid rain. He examines some of the myths associated with global commons concerns and anticipates the need for strategic interactions among nations to confront the problems.

Sarewitz, Daniel, and Roger Pielke, Jr. "Breaking the Global-Warming Gridlock." *The Atlantic Monthly*, vol. 286, July 2000, pp. 54–64. The authors analyze the international and domestic political gridlock that has resulted from opposing scientific, political, and financial resources that have been dedicated to the problem of global warming. While the arguments and debates continue, little is being done to deal with the root causes of global environmental degradation or the human suffering that often accompanies it.

Schubert, Charlotte. "Global Warming Debate Gets Hotter." *Science News*, vol. 159, June 16, 2001, p. 372. Schubert analyzes Bush's response to a report published by the National Academy of Sciences (NAS), in which a panel of distinguished scientists concluded that the IPCC's Third Assessment Report policy summary provided an "admirable summary of research" and that human activity "very likely" has caused the increase in global temperatures since 1900. The report acknowledged that uncertainties remain about the role of anthropogenic emissions because of gaps in knowledge about natural climate variations. Bush acknowledged the problem of global warming but emphasized the uncertainties mentioned in the report. He also announced plans for programs to improve climate predictions and create new technologies that would help monitor emissions and lead to the development of cleaner energy sources. However, he offered no plans to cut greenhouse emissions in the immediate future.

Schwarz, Adam. "Looking Back at Rio." *Far Eastern Economic Review*, vol. 156, October 28, 1993, pp. 48ff. One year after the Earth Summit in Brazil, Schwarz considers the problem of balancing the need for economic development against the costs of environmental deterioration, particularly in Asia. He discusses specific programs that have been put in place in several Asian nations to combat global warming, but concurs that in developing nations, food production and other critical issues have higher priority than the problem of climate change.

Sheridan, Kathryn. "Kyoto Protocol Ready for Ratification After Marrakesh." *British Medical Journal*, vol. 323, November 17, 2001, p. 1,146. This article covers the process of negotiating a final, legally binding version of the Kyoto Protocol during the United Nations climate talks in Marrakesh, Morocco, in the fall of 2001.

Skolnikoff, Eugene B. "The Role of Science in Policy: The Climate Change Debate in the United States." *Environment*, vol. 41, June 1999, pp. 16–20ff. Skolnikoff discusses how uncertainties about the nature of climate change can affect the role that scientific evidence plays in developing domestic and international policies to deal with global warming.

Stanglin, Douglas. "No Room for Zealots." *U.S. News & World Report*, vol. 123, December 22, 1997, p. 24. Stanglin reports that the Clinton administration portrayed the Kyoto Protocol as only a modest step toward ending global warming to avoid criticism of Vice President Al Gore as an environmental zealot whose policies would cost jobs. While polls suggested that the public would support a modest tax increase or rise in energy costs to develop cleaner, less expensive technologies, Republicans hoped that the quick passage of resolutions condemning the protocol would rally public support against it.

Sudetic, Chuck. "Bush's CO_2 Flip-Flop: The Surprising Truth." *Rolling Stone*, May 10, 2001, pp. 43–45. Sudetic covers the complex political wrangling behind George W. Bush's campaign promise to put mandatory limits on carbon dioxide emissions and his subsequent announcement, after taking office, that global warming science was too uncertain to warrant signing the Kyoto Protocol.

Sung, Juneho. "Rio Earth Summit." *Business Korea*, vol. 10, July 1992, pp. 34ff. Sung reports on the recently convened United Nations Conference on the Environment in Rio de Janeiro, Brazil. He analyzes what was accomplished at the summit and explores the tough choices and critical concerns that must be faced to fully address global environmental concerns.

Taylor, Jerry. "Clouds over Kyoto: The Debate over Global Warming." *Regulation*, vol. 21, Winter 1998, pp. 57–63. This article, published by the Cato Institute, argues that the debate over what to do about global warming has been driven more by politics than by science or economics. Because evidence for far-reaching disaster is scarce, Taylor argues that there is no compelling need to act immediately to curb carbon emissions.

Thompson, Dick. "Dubya's Next Showdown." *Time*, vol. 158, August 6, 2001, p. 24. Thompson reports on the global warming treaty discussions held in Bonn, Germany, where negotiators from nearly every country on Earth hammered out an agreement to limit greenhouse gas emissions. The exception was U.S. delegate Paula Dobriansky, who sat on the sidelines during negotiations and was heckled when she told the gathering that the Bush administration "will not abdicate our responsibility" to deal with global warming.

Victor, David G. "Climate of Doubt." *Sciences*, vol. 41, Spring 2001, pp. 18–23. This is one of many articles written in early and mid 2002 that prematurely predicted the death of the Kyoto Protocol. Victor, who wrote *The Collapse of the Kyoto Protocol* (2001), believes the treaty's architecture is flawed and will do nothing to stop global warming.

Vrolijk, Christiaan. "The Road from Marrakesh." *The World Today*, vol. 58, February 2002, pp. 26–27. Vrolijk analyzes the long-term prospects of the Kyoto Protocol as a tool to reduce greenhouse gas emissions following the international approval of the Marrakech Accords, which settled the outstanding issues of the protocol and resulted in the drafting of a detailed legal text.

Walker, Martin. "The Genoa Summit: What Did It Achieve?" *Europe*, September 2001, pp. S1–S2. Walker discusses the Group of 8 summit in Genoa, Italy, where leaders of the main industrialized countries held informal talks on various issues while demonstrators took to the streets to demand that attention be given to combating global warming and to debt relief, humanitarian aid, and health care for developing countries. During the resulting violence, one young demonstrator was shot dead by a police officer, hundreds were injured, and more than 220 demonstrators were arrested and jailed.

Wettestad, Jorgen. "Clearing the Air: Europe Tackles Transboundary Pollution." *Environment*, vol. 44, March 2002, pp. 32–40. This lengthy, footnoted article assesses progress made by the European Union and the Convention on Long-Range Transboundary Air Pollution in strengthening commitments to the control of air pollution in Europe. Because several of the targeted air pollutants are also greenhouse gases, these commitments also represent progress in the process of developing an effective international response to the threat of global warming. The article includes sidebars detailing the roles played by the European Union Commission and the Clean Air for Europe (CAFE) Program in mitigating air pollution.

Whitman, Christine Todd. "A Strong Climate Plan." *Time*, vol. 160, August 26, 2002, p. A48. Whitman, head of the U.S. Environmental Protection Agency (EPA), writes a spirited defense of President George W. Bush's controversial plan, unveiled in February 2002, to confront global warming.

Whitman, David. "The Hard Coal Facts: Bush Recants on Global Warming Pledge as King Coal Cheers and Big Green Jeers." *U.S. News & World Report*, vol. 130, March 26, 2001, pp. 16–18. After less than two months in office, President George W. Bush broke his campaign promise to support legislation that would require power plants to reduce emissions of carbon dioxide. He explained the policy shift by declaring, "We need a lot of coal" so Americans can "heat and cool their homes."

Wirth, Timothy. "Hot Air over Kyoto: The United States and the Politics of Global Warming." *Harvard International Review*, vol. 23, Winter 2002, pp. 72–77. Wirth, president of the United Nations Foundation and a former U.S. senator, recounts the events of 2001 regarding international global warming policy, including George W. Bush's rejection of the Kyoto Protocol and the rest of the world's subsequent determination to fulfill the treaty's promise without the participation of the United States.

Yachnin, Jennifer. "Bush Proposes New Programs in Climate Science." *The Chronicle of Higher Education*, vol. 47, June 22, 2001, p. A23. Following the release of a report on the status of climate-change science by the National Academy of Sciences (NAS), President Bush announced the creation of several new programs to study greenhouse gases. Among these were the National Climate Change Technology Initiative, which, under the auspices of the Energy Department and Commerce Department, would improve basic climate change research at universities and national laboratories, and the U.S. Climate Change Research Initiative, in which the secretary of commerce would recommend increased funding to "priority" global warming research areas. A primary focus of both programs would be the development of new greenhouse gas emissions monitoring technology and new energy technology. The NAS also recommended building a global climate-observation system and initiating a partnership with Japan and the European Union to create a climate-change model to explore the causes and impacts of global warming. Bush vowed to follow through with these suggestions as well.

Yokota, Yozo. "International Justice and the Global Environment." *Journal of International Affairs*, vol. 52, Spring 1999, pp. 583–598. Yokota analyzes the inability of traditional international law based on interstate relations to deal with global warming and other cross-border environmental threats. He suggests that effective means to prevent environmental damage must take into account interstate, interpersonal, and intergenerational justice.

INTERNET DOCUMENTS

Bush, George W. "President Announces Clear Skies and Global Climate Change Initiative." Presidents Council on Environmental Quality. Available online. URL: http://www.whitehouse.gov/news/releases/2002/02/20020214-5.html. February 14, 2002. This is the text of President Bush's speech in which he announced his domestic plan for dealing with global warming. The web page also provides access to a fact sheet summarizing the plan, as well as the Global Climate Change Policy Book.

Michaels, Patrick J. "Review of the *2001 U.S. Climate Action Report.*" Cato Institute Available online. URL: http://www.cato.org/research/articles/michaels0206.pdf. June 3, 2002. Michaels, a global warming skeptic, offers a point-by-point analysis of the *2001 U.S. Climate Action Report.* He finds the document "more balanced" than the IPCC Third Assessment Report but finds that some sections rely too heavily on the U.S. National Assessment (USNA) of climate change. The USNA, Michaels contends, is based on flawed climate models that "perform worse than a table of random numbers when applied to recent climate."

National Assessment Synthesis Team. "Climate Change Impacts on the United States: The Potential Consequences of Climate Variability and Change." U.S. Global Change Research Program. Available online. URL: http://www.usgcrp.gov/usgcrp/library/nationalassessment/overview.htm. Updated on March 30, 2003. This 2000 report, called for by the Global Change Research Act of 1990, identifies vulnerabilities and assesses potential impacts of global warming on each region of the United States.

United Nations. "United Nations Framework Convention on Climate Change." Available online. URL: http://unfccc.int/resource/conv/conv.html. Downloaded on April 2, 2003. 1992. This is the full text of the English-language version of the UNFCCC.

———. "Kyoto Protocol to the United Nations Framework Convention on Climate Change." Available online. URL: http://unfccc.int/resource/

docs/convkp/kpeng.html. Downloaded on April 2, 2003. 1997. This is the full text of the English-language version of the Kyoto Protocol.

U.S. Environmental Protection Agency. "U.S. Climate Action Report 2002: Third National Communication of the United States of America under the United Nations Framework Convention on Climate Change." Available online. URL: http://yosemite.epa.gov/OAR/globalwarming.nsf/uniquekeylookup/SHSU5BWHU6/$File/uscar.pdf. Posted in May 2002. This document represents the third formal national communication on climate change under the UNFCCC. It describes the national circumstances of the United States concerning global warming, identifies existing and planned policies and measures, indicates future trends in greenhouse gas emissions, outlines expected impacts and adaptation measures, and provides information on financial resources, technology transfer research, and systematic observations.

SCIENCE AND RESEARCH

BOOKS

Alley, Richard B. *The Two-Mile Time Machine: Ice Cores, Abrupt Climate Change, and Our Future.* Princeton, N.J.: Princeton University Press, 2000. During the 1990s, while studying ice cores drilled in Greenland, Alley and his colleagues discovered that the last ice age came to an end over a period of only a few years rather than gradually, as had been previously supposed. In this book, Alley recounts the history of ice coring and the details of his team's research, then provides an account of the Earth's wildly fluctuating climatic history. Finally, he warns that although the climate has been unusually mild and stable during the past 10,000 years, facilitating the rise of agriculture and industry, the planet could face another abrupt change at any time, quite possibly triggered by humans releasing massive amounts of carbon dioxide into the atmosphere.

Arnell, Nigel. *Global Warming, River Flows and Water Resources.* Ellis Horwood Series in Water and Wastewater Technology. New York: John Wiley and Sons, 1996. Arnell reviews the potential impacts of global warming on river flows and water resources and discusses the technical aspects of climate change impact assessments. He also reviews many of the published studies into possible changes in hydrology as a result of global warming.

Bailey, Ronald, ed. *Global Warming and Other Eco-Myths: How the Environmental Movement Uses False Science to Scare Us to Death.* Roseville, Calif.: Prima Publishing, 2002. Bailey, science correspondent for *Reason* magazine and adjunct scholar with the Competitive Enterprise Institute, has compiled a series of essays that seek to debunk scientific data and statistics used by environmentalists. Among the contributors is Nobel laureate Norman Borlaug. Topics covered include claims by environmentalists that Antarctica is melting because of global warming, that global population is growing faster than the ability to produce food, and that modern pesticides are increasing cancer rates.

Bigg, Grant R. *The Oceans and Climate.* New York: Cambridge University Press, 1996. This fairly technical book, suitable for undergraduates, looks at the complex interaction between the oceans and the atmosphere with an emphasis on the process of carbon dioxide storage and its role in global climate change. Numerous diagrams, charts, and maps help illustrate difficult-to-grasp theories. The book includes four appendices, a glossary, a bibliography, and an index.

Brandenburg, John E., and Monika Rix Paxson. *Dead Mars, Dying Earth.* Freedom, Calif.: Crossing Press, 1999. Physicist Brandenburg has developed a theory he calls Oxygen Inventory Depletion (OID), the idea that Earth's oxygen levels are falling because of increasing levels of carbon dioxide in the atmosphere. The processes that support life on Earth, the authors claim, are now being reversed by massive deforestation and by the use of fossil fuels and chemicals that are causing global warming and the expansion of the ozone hole. Using data culled from studies of Mars—including evidence that water once flowed on the surface and, more dubiously, photographs of the infamous "Face on Mars"—the authors theorize that the red planet once supported life and that Earth, if its oxygen levels fall to irreversibly low levels, will eventually become a dead planet as well.

Chambers, Frank, et al., eds. *Climate Change: Critical Concepts in Geography and Environment.* 4 vols. New York: Routledge, 2002. This four-volume collection of published papers seeks to update researchers on significant developments in the field of climate research. Areas of focus include evidence for natural variability of climate; the role of various forcing factors in climate change; arguments, both pro and con, concerning the "solar" and "anthropogenic" theories of climate change; and hypotheses about future climate.

Edgerton, Lynne T. *The Rising Tide: Global Warming and World Sea Levels.* Washington, D.C.: Island Press, 1991. Published under the auspices of the Natural Resources Defense Council, this book explores the major effects that global warming–induced sea level rise will have on coastal areas around the world. The author then analyzes federal and international policies that deal with climate change and recommends steps that should be taken at the federal and state level to reduce global warming's adverse effects. The reader will also find appendixes, notes, a bibliography, and an index.

Graedel, Thomas E., and Paul J. Crutzen. *Atmosphere, Climate, and Change.* New York: Scientific American Library, 1997. Yale University professor

Graedel and Nobel Prize–winning chemist Crutzen have written an excellent introduction to climate change, revealing both how climate fluctuations have affected humans and ecosystems around the globe and how humans have affected the climate. Colorful graphics and photographs enhance the clearly written text, and the list of further reading provides a guide for additional research.

Gutnik, Martin J. *Experiments That Explore the Greenhouse Effect.* Investigate! Series. Brookfield, Conn.: Millbrook Press, 1991. This brief book introduces the scientific method of investigation, then details a series of simple experiments that illustrate various concepts that are vital to the understanding of climate change. These include the physical and chemical properties of the atmosphere, air movements due to convection, temperature inversions, carbon dioxide's heat-holding capabilities, the self-sustaining nature of rain forests, the emission of sulfur dioxide from volcanoes, the effects of industrial carbon dioxide on plant growth, and the effects of global warming on human settlements. Included is a glossary, a brief bibliography, and an index.

Horrigan, John B., et al. *Taking a Byte out of Carbon: Electronics Innovations for Climate Protection.* Washington, D.C.: World Resources Institute, 1998. This book profiles 14 electronics and communications companies that have already developed innovative electronic technologies capable of reducing greenhouse gas emissions, even as policy makers continue to argue that emission abatement is not technologically or economically feasible.

Houghton, John. *Global Warming: The Complete Briefing.* 2d ed. Cambridge, England: Cambridge University Press, 1997. Although fairly technical, this book will be of interest to anyone who seeks a deeper understanding of the science upon which the Intergovernmental Panel on Climate Change bases its assessment reports. Houghton, cochairman of the IPCC's Scientific Assessment Working Group, offers chapters on the greenhouse effect, carbon dioxide and the carbon cycle, feedbacks in the climate system, projections of global average temperature, energy conservation, the impact of global warming on sea level and freshwater resources, and more. Each chapter ends with a summary, a list of questions for discussion, and notes. The book includes a glossary, an index, and a large number of charts, diagrams, and illustrations.

Hoyt, Douglas V., and Kenneth H. Schatten. *The Role of the Sun in Climate Change.* New York: Oxford University Press, 1997. The authors discuss the theory that some, or even all, of the Earth's warming trend can be attributed to natural variations in the Sun's output of radiation. Attention is given to historical studies of solar influences on climate and how climatologists combine thousands of local observations to create a picture of past and present global climate. Included is a chapter on alternative climate change theories (the volcanic hypothesis, increased anthropogenic aerosols, and greenhouse gas warming); glossaries of solar, astronomical, and mathematical terms; solar and terrestrial data charts; and an extensive bibliography.

International Energy Agency and the Organization for Economic Cooperation and Development. *Development and Deployment of Technologies to Respond to Global Climate Change Concerns.* Paris: OECD/IEA, 1994. This report consists of the proceedings of a 1994 conference convened by the International Energy Agency and the Organization for Economic Cooperation and Development to pinpoint strategies aimed at enhancing the development of new technologies that respond to climate change concerns. Discussions focus on ways to encourage international collaboration and cooperation between industry and nongovernmental organizations.

Jarvis, P. G., and Anne M. Aitkin. *European Forests and Global Change: The Likely Impacts of Rising* CO_2 *and Temperature*. Cambridge, England: Cambridge University Press, 1998. The authors describe the results of research in which major European tree species are subject to experimentally manipulated environmental conditions to determine how global warming affects photosynthesis, respiration, and development. Data collected from these experiments are used to generate models of the likely response of European forests to the predicted changes in climate.

Mayewski, Paul Andrew, and Frank White. *The Ice Chronicles: The Quest to Understand Global Climate Change.* Hanover, N.H.: University Press of New England, 2002. This book documents climate change research scientist Mayewski's efforts to lead the National Science Foundation's Greenland Ice Sheet Project Two (GISP2). This expedition, in which 25 universities took part, involved traveling to Antarctica, Asia, and Greenland to collect ice cores that hold clues to climate history dating back 100,000 years. The new techniques developed by Mayewski's team revealed major events of the past, the dramatic effects humans have had on the chemistry of the atmosphere, and the rapid and dramatic changes that the Earth's climate undergoes every few thousand years.

Michaels, Patrick J. *Sound and Fury: The Science and Politics of Global Warming*. Washington, D.C.: Cato Institute, 1992. Global warming skeptic Michaels, professor of environmental sciences at the University of Virginia, questions the popular notion that most scientists subscribe to the belief that climate change poses an apocalyptic threat to humans. Along the way, Michaels argues that slight warming will have such benign effects as extending agricultural growing seasons, and that higher concentrations of carbon dioxide, far from acting as a pollutant, will help plants grow and will therefore supply the world with cheaper, more plentiful food.

Michaels, Patrick J., and Robert C. Balling, Jr. *The Satanic Gases: Clearing the Air About Global Warming*. Washington, D.C.: Cato Institute, 2000.

Michaels and Balling believe that global warming is vastly overrated as a threat to Earth's environment, criticizing the conclusions of scientists (particularly those involved with the Intergovernmental Panel on Climate Change) who issue dire warnings about imminent disaster. Rather than claim that climate change is a complete myth, they argue that temperature rise during the 21st century will be no greater than that experienced during the last third of the 20th century, a time during which food supplies, wealth, and life expectancy all improved on a scale never seen before. Furthermore, any additional heating will occur during the winter, reducing the potentially deadly effects of frigid temperatures. The book contains many charts, diagrams, and maps, as well as a list of references.

Miller, Clark A., and Paul N. Edwards, eds. *Changing the Atmosphere: Expert Knowledge and Environmental Governance.* Politics, Science, and Environment Series. Cambridge, Mass.: MIT Press, 2001. Experts from a variety of fields, including biology, information studies, public policy, and philosophy, explore the link between the scientific understanding of climate change and the way that relevant public policy decisions are made by politicians and government officials. Topics include computer models of the climate, weather modification, environmental justice, and the complex relation between international cooperation among scientists and U.S. foreign policy. The list of references is extensive.

Nance, John *J. What Goes Up: The Global Assault on Our Atmosphere.* New York: William Morrow, 1991. Science and technology writer Nance focuses on the lives of the scientists who conduct research into global warming and ozone issues. He reveals how scientists who warn about threats to the environment are often vilified by the chemical and oil companies whose products and profits are threatened by the revelations and how these companies attempt to suppress and discredit both the researchers and their findings. Appendixes provide additional information on the scientists discussed in the text, as well as references for further research.

Nilsson, Annika. *Greenhouse Earth.* New York: John Wiley and Sons, 1992. Published on behalf of the Scientific Committee on Problems of the Environment (SCOPE), the International Council of Scientific Unions (ICSU), and the United Nations Environment Programme (UNEP), this book provides a detailed view of the state of climate change research just before the 1992 Earth Summit in Brazil. Diagrams and charts illustrate concepts discussed in the IPCC's 1990 assessment of climate change, the book's primary source of information.

Peters, Robert L., and Thomas E. Lovejoy, eds. *Global Warming and Biological Diversity.* New Haven, Conn.: Yale University Press, 1992. In this collection of essays, distinguished scientists discuss the potential for global warming to disrupt natural ecosystems and even cause species extinction. The book is divided into five sections, beginning with an overview of global warming and followed by responses of ecosystems to past climate change, general ecological and physiological responses to climate change, studies of specific regions and sites, and implications for conservation. The articles are somewhat technical at times but contain essential information for researchers specifically interested in the subject of biological diversity.

Philander, S. George. *Is the Temperature Rising? The Uncertain Science of Global Warming.* Princeton, N.J.: Princeton University Press, 2000. Philander, a geoscientist at Princeton University, presents in clear language evidence that the proliferation of greenhouse gases and other pollutants in the atmosphere is having damaging effects on the planet. Although he believes the Earth will regulate itself over the long run, he worries that short-term changes in climate may be drastic and prove impossible to accurately predict. Philander explains, however, that the inability to predict the future is no reason to avoid taking action to fix the problem, and he strongly disagrees with skeptics who think mitigation strategies should be implemented only after researchers present more accurate scientific information about global warming. In fact, he argues, the uncertainty is precisely why policy makers should act to avoid a potentially calamitous future.

Wood, Chris M., ed. *Global Warming: Implications for Freshwater and Marine Fish.* Cambridge, England: Cambridge University Press, 1997. The 15 papers in this book focus on the effects of water temperature on fish, most of which have no physiological ability to regulate their body temperature. Discussions include temperature thresholds for protein adaptation, wild Atlantic salmon stocks in Europe, thermal stress and muscle function, cardiovascular performance, reproduction, embryonic and larval development, cod stocks, growth rate, interactive effects of temperature and pollutant stress, behavioral compensation for long-term thermal change, and the thermal niche of fishes. Emphasis in many chapters is given to speculations about the long-term physiological and ecological implications to fish of a 2- to 4-degree-Celsius (3.6- to 7.2-degree Fahrenheit) global warming scenario over the first half of the 21st century.

Woodwell, George M., and Fred T. MacKenzie. *Biotic Feedbacks in the Global Climate System: Will the Warming Feed the Warming?* New York: Oxford University Press, 1995. The essays in this book were originally presented during a conference on biotic feedbacks at the Woods Hole Research Center in Massachusetts in October 1992, four months after the Earth Summit in Brazil. Experts in agriculture, botany, oceanography, chemistry, geography, environmental sciences, and atmospheric sciences from around the world discuss the possible effects of global warming on vegetation, including the possibility that rising temperatures will change the distribution of forests and affect the ability of plants to metabolize carbon dioxide, adding further to the greenhouse effect (known as positive feedback). The essays are extremely technical and will be most helpful to those interested in delving deeply into complex scientific aspects of biotic feedbacks.

ARTICLES

"Air Pollution Control Efforts Could Add to Global Warming." *Bulletin of the American Meteorological Society*, vol. 82, July 2001, pp. 1,451ff. Climate researchers warn that efforts to reduce emissions of human-made nitrogen oxides with the goal of controlling ozone levels in the lower atmosphere would result in increased methane emissions, thereby adding to the problem of global warming.

"Air Repair." *Omni*, vol. 15, June 1993, p. 62. This article examines the theory that the best way to fight global warming and do away with ozone holes is through the use of geoengineering, or intentionally tampering with the environment. Among the proposals are reforesting more than 28 million hectares (about 17,500 square miles) of farmland so trees can absorb more carbon dioxide; sending giant mirrors into space to reflect excess sunlight away from the planet; using naval guns, rockets, or balloons to carry dust into the stratosphere to block sunlight; fertilizing the ocean to promote the growth of carbon dioxide–absorbing algae; and injecting hydrocarbons into the Antarctic stratosphere to react with chlorofluorocarbons (CFCs) before they can destroy the ozone layer.

Bains, Santo, Richard M. Corfield, and Richard D. Norris. "Mechanisms of Climate Warming at the End of the Paleocene." *Science*, vol. 285, July 30, 1999, pp. 724–727. Oxygen and carbon isotope records indicate that an abrupt episode of global warming that marked the end of the Paleocene epoch (about 55.5 million years ago) may have been caused by the degassing of biogenic methane hydrate.

Barron, Tom. "CIA Archives, Spysats Tackle New Mission." *Environment Today*, vol. 3, November 1992, pp. 3ff. Brown reports on proposals, resisted by Pentagon officials but supported by the scientific community, to use existing intelligence data and collection capability for environmental research. Scientists are hoping to use data collected by spy satellites, ships, submarines, and reconnaissance aircraft over a 30-year period of cold war espionage to analyze measurements of atmospheric and other ecological conditions for information about global warming, polar ice thickness, sea temperatures, deforestation, and desertification.

Annotated Bibliography

Broecker, Wallace S. "Glaciers That Speak in Tongues and Other Tales of Global Warming." *Natural History*, vol. 110, October 2001, pp. 60–69. Broecker discusses how the study of past records of glacial advances and retreats enables paleoclimatologists to reconstruct temperature cycles during the past several thousand years. This data provides a baseline from which researchers can attempt to determine the effect that human activity is having on the planet's climate systems.

———. "Was the Medieval Warm Period Global?" *Science*, vol. 291, February 23, 2001, pp. 1,497–1,499. Broecker argues that the Medieval Warm Period, during which the Vikings colonized Greenland, was a global rather than regional climate event. He also believes that the warming that occurred at the beginning of the Industrial Revolution, before the human-induced emission of substantial amounts of greenhouse gases, was the most recent in a series of similar temperature upswings spaced at roughly 1,500-year intervals throughout the Holocene.

Brown, Stuart F. "The Sound of Global Warming." *Popular Science*, vol. 247, July 1995, p. 59. This article examines the Scripps Institution of Oceanography's acoustic thermometry of ocean climate (ATOC) project, which uses sound waves to measure ocean temperatures with the goal of developing better global warming models. Scientists worry, however, that the sound waves may interfere with the activities of whales, sea lions, and other sea life.

Bunyard, Peter. "How Climate Change Could Spiral Out of Control." *The Ecologist*, vol. 29, March/April 1999, pp. 68–74. Bunyard examines positive feedbacks involving living organisms that may destabilize the climate system and accelerate the process of global warming. Among these is the impact of ocean warming on populations of phytoplankton, which in turn will affect carbon dioxide absorption. Bunyard claims that existing climate models, because they fail to take such positive feedbacks into account, may be underestimating the potential impact of global warming.

"Cooling Trend in Antarctica." *The Futurist*, vol. 36, May/June 2002, p. 15. While the average global air temperature increased 0.6 degree Celsius (1 degree Fahrenheit) during the 20th century, researchers at the U.S. National Science Foundation report that areas of Antarctica experienced a cooling trend of 0.7 degree Celsius (1.3 degrees Fahrenheit) per decade between 1986 and 2000.

Crutzen, Paul J., and Veerabhadran Ramanathan. "The Ascent of Atmospheric Sciences." *Science*, vol. 290, October 13, 2000, pp. 299–304. Pioneering atmosphere researchers Crutzen and Ramanathan recount developments in chemistry and meteorology before the 1970s that led to the development of the atmospheric sciences and the realization that humans have influenced atmospheric ozone levels and global climate.

D'Agnese, Joseph. "Why Has Our Weather Gone Wild?" *Discover,* vol. 21, June 2000, pp. 72–78. D'Agnese discusses scientific evidence that global warming is responsible for a seeming increase in extreme weather conditions, including droughts and heavier rainfall.

Delworth, Thomas L., and Thomas R. Knutson. "Simulation of Early 20th Century Global Warming." *Science,* vol. 287, March 24, 2000, pp. 2,246–2,250. Global warming during the 20th century occurred primarily in two distinct 20-year periods, from 1925 to 1944 and from 1978 to 2000. Most climate scientists agree that the warming of the latter period can be attributed to a human-induced increase of greenhouse gases. The earlier period, however, preceded the era of rapid increases in greenhouse gas emissions. The authors consider other possible causes of global temperature rise during this time.

"Earth's Becoming a Greener Greenhouse." *Bulletin of the American Meteorological Society,* vol. 82, November 2001, pp. 2,482–2,486. Data collected from NASA satellites has shown that since 1981, regions of the Northern Hemisphere above 40 degrees North latitude have experienced more vigorous plant growth than before. Scientists theorize that this phenomenon is the result of rising temperatures, possibly linked to steadily rising levels of greenhouse gases in the atmosphere. In Eurasia, the growing season became 18 days longer between 1981 and 2001, while the North American growing season expanded by 12 days over the same period.

Easterling, David R., et al. "Observed Variability and Trends in Extreme Climate Events: A Brief Review." *Bulletin of the American Meteorological Society,* vol. 81, March 2000, pp. 417–425. Also available online. (URL: http://lwf.ncdc.noaa.gov/oa/pub/data/special/extr-bams2.pdf). As a result of increasing economic losses and death tolls, interest in extreme weather events has increased in recent years. The authors report on the difficulty of determining trends in extreme weather phenomena, mostly due to the lack of high-quality long-term data.

Epstein, Paul R. "Is Global Warming Harmful to Health?" *Scientific American,* vol. 283, August 2000, pp. 50–57. Epstein reports on computer model predictions that many diseases will increase as global temperatures rise. He also examines the dramatic rise in the number of malaria and cholera cases that has already accompanied warmer temperatures and increased flooding in tropical regions.

Fedorov, Alexey V., et al. "Is El Niño Changing?" *Science,* vol. 288, June 16, 2000, pp. 1,997–2,002. Researchers consider the question of whether global warming has affected or will affect El Niño in light of recent advances in observational and theoretical studies of the phenomenon.

Fetter, Steve. "Energy 2050." *Bulletin of the Atomic Scientists,* vol. 56, July/August 2000, pp. 28–38. Fetter discusses the replacement of fossil fuels with energy sources that emit little or no carbon dioxide. Among the alternative energy sources discussed are nuclear fission, biomass fuels, solar and wind energies, and decarbonized fossil fuels.

Gantenbein, Douglas. "Harvard Expert Debunks Global Warming 'Models.'" *Consumers' Research Magazine*, vol. 84, May 2001, pp. 20–23. In this interview, Sallie Baliunas of the Harvard-Smithsonian Center for Astrophysics discusses the uncertainty of global warming science and the often-overlooked possibility that the Sun is responsible for much of the climate's fluctuation.

———. "The Heat Is On." *Popular Science*, vol. 255, August 1999, pp. 54–59. This article profiles Jerry Mahlman, director of the National Oceanic and Atmospheric Administration's Geophysical Fluid Dynamics Laboratory at Princeton University in New Jersey. He is among the climate scientists convinced that global warming is well under way and that industrial output of greenhouse gases is responsible for it.

Greenstone, Matthew H. "Greenhouse Gas Mitigation: The Biology of Carbon Sequestration." *Bioscience*, vol. 52, April 2002, p. 323. Among the proposals that scientists have made to remove carbon dioxide from the atmosphere is the injection of the gas into the deep sea. Greenstone discusses the effect this might have on deep-sea life, in the end suggesting that greater fuel efficiency and the replacement of fossil fuels with renewable energy sources may be a better idea.

Hansen, James E., et al. "Global Climate Data and Models: A Reconciliation." *Science*, vol. 281, August 14, 1998, pp. 930–932. Hansen and his colleagues discuss the implications of a discovery that satellite observations of a cooling trend in the troposphere, which were at odds with surface readings and climate models, were based on erroneous data. The corrected reading, according to the scientists, brings the temperature profile in close accord with published climate model simulations.

Hayashi, Alden M. "Winds of Change." *Scientific American*, vol. 282, January 2000, pp. 40–41. Hayashi reports on a growing concern among some scientists that global warming could lead to an increase in hurricane intensity. He analyzes the possibility that these stronger hurricanes, if they were to hit a major metropolitan area, would cause great damage to skyscrapers.

Houghton, John. "The Science of Global Warming." *Interdisciplinary Science Reviews*, vol. 26, Winter 2001, pp. 247–257. Houghton, cochair of the Science Assessment Working Group of the Intergovernmental Panel on Climate Change, provides a general overview of the greenhouse effect and discusses in detail the "strong scientific evidence" that the average global temperature is rising because of anthropogenic fossil fuel emissions.

Johannessen, Ola M., et al. "Satellite Evidence for an Arctic Sea Ice Cover in Transformation." *Science*, vol. 286, December 3, 1999, pp. 1,937–1,939. Johannessen and colleagues discuss how microwave satellite remote sensing data has been used to show that the extent of the Arctic sea ice cover has decreased by about 3 percent per decade in since 1978.

Kaiser, Jocelyn. "Global Warming, Insects Take the Stage at Snowbird." *Science*, vol. 289, September 22, 2000, pp. 2,031–2,032. At the Ecological Society of America's 85th annual meeting, held in Snowbird, Utah, some 2,600 ecologists discussed topics ranging from ancient droughts and photosynthesis beneath snow to global warming to how trees resist insects.

Kerr, Richard A. "Ozone Loss, Greenhouse Gases Linked." *Science*, vol. 280, April 10, 1998, p. 202. Although many people have long confused global warming with stratospheric ozone destruction, a definitive connection between the two has never been made. Kerr reports on a computer model suggesting that greenhouse gases and chlorofluorocarbons may be working together to destroy ozone.

———. "Rising Global Temperature, Rising Uncertainty." *Science*, vol. 292, April 13, 2001, pp. 192–194. Despite acknowledging that climate prediction is fraught with uncertainties that may never be completely resolved, many scientists in the field remain adamant that the data indicates enough of a threat that action should be taken to fight global warming.

———. "The Tropics Return to the Climate System." *Science*, vol. 292, April 27, 2001, pp. 660–661. Recent research by paleontologists suggests that the tropical oceans, and especially the tropical Pacific, play a much more important role in long-term climate change than previously thought.

———. "Warming's Unpleasant Surprise: Shivering in the Greenhouse?" *Science*, vol. 281, July 10, 1998, pp. 156–158. Kerr discusses growing scientific evidence that the oceans control the Earth's climate and reports on a warning by Wallace Broecker, a marine geochemist at Columbia University, that global warming could shut down the global ocean current that warms the Northern Hemisphere.

Kowalok, Michael E. "Common Threads: Research Lessons from Acid Rain, Ozone Depletion, and Global Warming." *Environment*, vol. 35, July 1993, pp. 12ff. Kowalok provides an overview of research into the issues of acid rain, stratospheric ozone depletion, and global warming. He discusses how individual scientists interpret the results of research efforts to identify environmental threats, and how this data is used to develop programs to deal with these problems.

Krajick, Kevin. "Arctic Life, on Thin Ice." *Science*, vol. 291, January 19, 2001, pp. 424–425. Krajick reports that changing climatic conditions since 1978 have led to a 15 percent reduction in the area covered by sea ice in the Arctic. As a result, many animal and plant species that rely on the ice are suffering.

Levitus, Sydney, et al. "Warming of the World Ocean." *Science*, vol. 287, March 24, 2000, pp. 2,225–2,229. This technical article documents a rise in the mean temperature of the world ocean of 0.06 degree Celsius (0.11 degree Fahrenheit) between the mid-1950s and mid-1990s.

Liang, Stephen, Kenneth J. Linthicum, and Joel C. Gaydos. "Climate Change and the Monitoring of Vector-Borne Disease." *The Journal of the American Medical Association*, vol. 287, May 1, 2002, pp. 2,286ff. Many scientists believe that global warming has caused an increase in mosquito and tick reproduction and biting, which has led, in turn, to increases in the transmission of such insect-borne pathogens as Rift Valley fever, malaria, and Lyme disease. The authors of this article report how remote sensing technology and geographic information systems have been used to collect and consolidate environmental data relevant to the understanding of the relationship between climate and vector-borne disease. Such information can help health professionals prepare to deal with changes in the distribution of important infectious pathogens.

Martens, Pim. "How Will Climate Change Affect Human Health?" *American Scientist*, vol. 87, November/December 1999, pp. 534–541. Martens explores the consequences of global warming for public health issues such as reduced air quality, heat stress, the spread of infectious diseases, and rising sea level.

McCarthy, Michael. "Uncertain Impact of Global Warming on Disease." *The Lancet*, vol. 357, April 14, 2001, p. 1,183. A report by the U.S. National Academies of Science National Research Council (NRC) challenges widespread claims that global warming will cause an increase of malaria, dengue, yellow fever, and other infectious diseases. The report states that there is "little solid scientific evidence to support such conclusions."

MacDonald, Alexander E. "The Wild Card in the Climate Change Debate." *Issues in Science and Technology*, vol. 17, Summer 2001, pp. 51–56. MacDonald, director of the National Oceanic and Atmospheric Administration's Forecast Systems Laboratory in Boulder, Colorado, writes that one of the biggest and most often ignored threats of global warming is the possibility that it will trigger abrupt climate changes on a regional scale that could have long-lasting and devastating effects. He believes that the United States must develop a set of programs to improve its ability to predict regional climate change. Among his suggestions is supplementing data-collecting satellites, which provide an overall view of the Earth, with a more comprehensive system of local sensors that provide calibration and local detail. The government, which has shown its willingness to invest money in technology, should increase its spending on scientists themselves, bringing in more experts to conduct research on regional climate prediction. Above all, MacDonald continues, the United States needs a directed program of research focusing on regional climate change by identifying a government organization that would have comprehensive, overall responsibility for long-term climate prediction.

McDonald, Kim A. "An Ecological Network." *The Chronicle of Higher Education*, vol. 41, April 14, 1995, p. A8. McDonald discusses the details of experiments conducted by biologists who are studying long-term environmental changes in specific ecosystems to assess the impact of global warming.

McElwain, J. C., et al. "Fossil Plants and Global Warming at the Triassic-Jurassic Boundary." *Science*, vol. 285, August 27, 1999, pp. 1,386–1,390. A group of scientists analyze paleobotanical evidence indicating that a mass animal extinction at the Triassic-Jurassic boundary (about 200 million years ago) was accompanied by a fourfold increase in atmospheric carbon dioxide concentration and an associated 3- to 4-degree-Celsius (5.4- to 7.2-degree-Fahrenheit) greenhouse warming.

Perkins, Sid. "Antarctic Sediments Muddy Climate Debate." *Science News*, vol. 160, September 8, 2001, p. 150. Scientists who have studied sediments collected from portions of the ocean floor in the Antarctic that had recently been covered by ice shelves suggest that those shelves were only 2,000 years old. This has led some scientists to theorize that the area was free from ice for at least part of the year from about 2,000 to 5,000 years ago, suggesting that the disappearance of at least some of the Antarctic ice may be the result of natural cycles rather than human-caused global warming.

———. "Big Bergs Ahoy!: An Armada of Ice Sets Sail for the New Millennium." *Science News*, vol. 159, May 12, 2001, pp. 298–300. Ice shelves in the Antarctic have been shedding huge icebergs at an unprecedented rate. Although global warming may seem a likely explanation, researchers have been unable to definitively link the breakup of ice shelves to warmer temperatures. A sidebar discusses how a grounded iceberg near the Ross Island threatens thousands of penguins.

———. "Global Warming to Boost Cotton Yields." *Science News*, vol. 161, January 5, 2002, p. 15. Perkins reports on research conducted by Linda O. Mearns and Ruth Doherty—climatologists at the National Center for Atmospheric Research in Boulder, Colorado—showing that a doubling of atmospheric carbon dioxide will lead to bigger cotton harvests. The researchers attribute this prospective yield increase to the fertilizing effect of increased carbon dioxide and longer growing seasons caused by warmer temperatures.

———. "Satellites Verify Greenhouse-Gas Effects." *Science News*, vol. 159, March 17, 2001, p. 165. Perkins reports that data collected from satellites over 25 years support other evidence that greenhouse gases had an increasingly significant effect on the Earth's atmosphere during the final decades of the 20th century.

———. "Spring Forward." *Science News*, vol. 163, March 8, 2003, pp. 152–154. Perkins reports that scientists who study the response of organisms to seasonal and climatic changes have noted that the annual cycles for many animals are beginning earlier on average as global temperatures rise. Meanwhile, many plants and animals that prefer warmer temperatures have taken advantage of the warming climate to expand their ranges toward the North and South Poles or upslope, toward higher elevations.

Perritano, John. "Can We Stop Global Warming?" *Current Science*, vol. 85, December 17, 1999, pp. 8–9. This article examines strategies to fight global warming, including reducing use of fossil fuels, carbon sequestration, and improving Earth's albedo.

Pitelka, Louis F. "Plant Migration and Climate Change." *American Scientist*, vol. 85, September/October 1997, pp. 464–473. In this long essay, the product of a workshop on plant migration and climate change held in Australia in October 1996, Pitelka examines current understanding of plant migration based on the fossil record, contemporary invasions of nonnative plant species, and the effects of landscape patterning. He discusses how plants forced to migrate because of climate change must move through a landscape that has been degraded and rendered impassable by human activity, which may lead to extinction. Both regional problems and global consequences are considered.

Roach, Mary. "Antarctica's Hot Spot." *Discover*, vol. 20, November 1999, pp. 102–109. Roach offers a fascinating firsthand account of her journey to Lallemand Fjord, on the west coast of the Antarctic Peninsula, aboard the *Nathaniel B. Palmer*. Scientists aboard this research vessel braved gale-force winds and 12-meter (40-foot) waves to find out why one of the coldest places on Earth is heating up faster than anywhere else.

Robertson, G. Philip, et al. "Greenhouse Gases in Intensive Agriculture: Contributions of Individual Gases to the Radiative Forcing of the Atmosphere." *Science*, vol. 289, September 15, 2000, pp. 1,922–1,925. Robertson and his colleagues discuss the results of research designed to help determine the role that agriculture plays in the global fluxes of the greenhouse gases carbon dioxide, nitrous oxide, and methane.

Robinson, Peter J. "On the Definition of a Heat Wave." *Journal of Applied Meteorology*, vol. 40, April 2001, pp. 762–775. Heat waves are a major cause of weather-related deaths, and concerns are growing that global warming will increase the frequency, severity, duration, and range of severe heat waves. To help assess the danger, Robinson proposes a set of definitions based on the criteria for heat stress forecasts developed by the

National Weather Service (NWS). The lengthy article includes diagrams, tables, and a list of references.

Runyan, Curtis. "Ocean Warming Studies Bolster Evidence of Human Hand in Climate Change." *World Watch*, vol. 14, July/August 2001, p. 10. Two independent studies published in April 2001 showed a close match between computer projections of climate change and data showing the Earth's oceans have warmed 0.06 degree Celsius (0.11 degree Fahrenheit) in the past 40 years. Runyan considers this some of the most convincing evidence yet that anthropogenic emissions are playing a significant role in the warming of the Earth.

Satchell, Michael. "A Sickly Green? Global Warming." *U.S. News & World Report*, vol. 131, September 17, 2001, p. 86. Satchell reports on the finding, based on data collected by satellites, that the growth of vegetation in the Northern Hemisphere above 40 degrees latitude has increased dramatically since 1981. Many scientists believe that the longer, more vigorous growing season is the result of rising temperatures most likely caused by increasing levels of atmospheric carbon dioxide.

"Satellites Shed Light on a Warmer World." *Bulletin of the American Meteorological Society*, vol. 83, March 2002, p. 346. This article provides an account of efforts by James Hansen of NASA's Goddard Institute for Space Studies and Marc Imhoff of NASA's Goddard Space Flight Center to use satellite images of nighttime lights to identify weather stations where urbanization will have the least impact on temperature data. Using data from these remote stations, the researchers were able to compile a climate record largely free of influence from the urban heat island effect, to which some global warming skeptics ascribe rising temperatures.

Shaver, Gaius R., et al. "Global Warming and Terrestrial Ecosystems: A Conceptual Framework for Analysis." *Bioscience*, vol. 50, October 2000, pp. 871–882. A group of scientists analyzes efforts to predict the complex and varied ways that various ecosystems will respond to global warming over the next several decades.

Shepherd, Andrew, et al. "Inland Thinning of Pine Island Glacier, West Antarctica." *Science*, vol. 291, February 2, 2001, pp. 862–864. This article reports on the melting of the Pine Island Glacier and the possibility that this could result in increased ice discharge from the West Antarctic Ice Sheet. If the West Antarctic Ice Sheet were to melt into the ocean, global sea levels could rise substantially.

"Sick of the Weather." *Current Science*, vol. 86, January 19, 2001, pp. 12–13. This article examines the link between public health problems such as malaria, cholera, heart attack, stroke and asthma, and extreme weather conditions.

Simpson, Sarah. "Melting Away." *Scientific American*, vol. 282, January 2000, pp. 19–20. This article examines geologic evidence and satellite images indicating that massive ice shelves in West Antarctica are melting, which may cause the sea level to rise.

———. "Methane Fever." *Scientific American*, vol. 282, February 2000, pp. 24–27. Simpson discusses a series of undersea methane explosions 55 million years ago that may have emitted greenhouse gases into the atmosphere at a rate similar to that at which carbon dioxide is being pumped into the air today. Some scientists believe this event may have caused the most rapid warming episode of the past 90 million years.

Slotnick, Rebecca Sloan. "Red All Over." *American Scientist*, vol. 90, January/February 2002, p. 31. Slotnick discusses global climate change maps that show areas of more rapid warming in red. Scientists use these to illustrate changes in the Earth's surface-air temperatures from the year 2000 relative to the global mean temperature from 1951 to 1980.

Solomon, Burt. "A Radical Approach to Global Warming." *National Journal*, vol. 31, October 2, 1999, pp. 2,828–2,829. Solomon examines efforts by scientists and engineers to solve the problem of global warming by conducting research into carbon sequestration—ways to remove carbon and other pollutants from the air and store them so they cannot trap the Earth's heat. Ideas range from the low-tech option of planting new forests to injecting carbon dioxide into underground geologic formations to experimenting with "artificial photosynthesis" by engineering microbes or other molecular devices to boost plants' consumption of greenhouse gases.

"Tales from the Ice." *Science World*, vol. 12, November 12, 2001, pp. 20–22. This introductory article discusses the work of scientists who travel to Antarctica to collect ice cores, which can be studied for clues about the Earth's past climates. Fascinating sidebars illustrate the kind of information that has appeared in ice samples from 1988 (elevated radiation levels from the Chernobyl nuclear power station explosion), 1900 (elevated carbon dioxide and methane levels from increased burning of fossil fuels), and 73,000 B.C.E. (elevated levels of ash from the eruption of Toba, an ancient volcano in Indonesia).

Taubes, Gary. "Apocalypse Not." *Science*, vol. 278, November 7, 1997, pp. 1,004–1,006. According to many infectious-disease specialists, predictions that global warming will spark epidemics of malaria, dengue fever, yellow fever, cholera, hantavirus, Ebola, and other diseases have little basis. Even if such diseases begin to increase under the influence of rising temperatures, public health measures will be taken to deal successfully with the problem.

Uppenbrink, Julia. "Arrhenius and Global Warming." *Science*, vol. 272, May 24, 1996, p. 1,122. This article examines the research of Swedish scientist

Svante August Arrhenius, the first person to link variations in the concentration of carbon dioxide in the atmosphere to changes in climate.

Uttal, Taneil, et al. "Surface Heat Budget of the Arctic Ocean." *Bulletin of the American Meteorological Society*, vol. 83, February 2002, pp. 255–275. This lengthy essay details the findings of the Surface Heat Budget of the Arctic Ocean (SHEBA), a year-long research camp centered on a Canadian icebreaker frozen in the Arctic ice pack. Scientists collected atmospheric, oceanographic, and cryospheric data to document, understand, and predict significant changes in the Arctic that are thought to result from a combination of natural modes of variability and anthropogenic greenhouse warming. The article includes a list of references.

Vinnikov, Konstantin Y., et al. "Global Warming and Northern Hemisphere Sea Ice Extent." *Science*, vol. 286, December 3, 1999, pp. 1,934–1,937. Data collected from surface monitors and satellites show a decrease in Northern Hemisphere sea ice during the past 46 years. Researchers have developed two models that predict continued decreases in sea ice thickness and extent throughout the 21st century.

Waller, J. Michael, Wade-Hahn Chan, and Daniel George. "Cold Facts About Global Warming." *Insight on the News*, vol. 18, September 23, 2002, p. 6. The authors question the validity of global warming theory, based on a report from NASA that satellite observations have determined that during the last 20 years the ice in some regions of the Antarctic has increased rather than melted. Two separate studies issued by the American Meteorological Society also found increases in Antarctic ice between 1987 and 1996, while a 2001 study published in the journal *Geophysical Research Letters* found little change in ice mass during the 1990s.

Watson, Robert T., and David E. Wojick. "Do Scientists Have Compelling Evidence of Global Warming?" *Insight on the News*, vol. 17, March 12, 2001, pp. 40–43. Robert Watson, chairman of the IPCC at the time this article was written, argues that the 2001 IPCC report provides strong evidence for human influence on the Earth's climate system.

Wilinson, Todd. "The Woman Behind the Exhaust-Free Car: A GM Engineer Is Eyeing a Hydrogen-Fueled Car That Would Replace the Internal-Combustion Engine." *Christian Science Monitor*, April 21, 2003, pp. 3ff. The author profiles Christine Sloane, a senior engineer at General Motors, who is working to design commercially viable hydrogen cars.

INTERNET DOCUMENTS

Marland, Gregg, and Tom Boden. "The Increasing Concentration of Atmospheric Carbon Dioxide: How Much, When, and Why?" Carbon Dioxide Information Analysis Center. Available online. URL:

http://cdiac.esd.ornl.gov/epubs/other/Sicilypaper.pdf. Posted on August 2001. This paper reviews the scientific evidence in support of the theory that levels of atmospheric carbon dioxide are increasing and this increase is due largely to the combustion of fossil fuels. The authors also look at scenarios for the future use of fossil fuels and what these may mean for the future of atmospheric chemistry.

Mazza, Patrick, and Rhys Roth. "Global Warming Is Here: The Scientific Evidence." Climate Solutions. Available online. URL: http://www.climatesolutions.org/pubs/gwih.html. Posted on February 2002. The authors provide an overview of the greenhouse effect and the data that has been collected to support theories of global warming.

National Oceanic and Atmospheric Administration. "Strategy for Climate Change Research Defined at Science Workshop. Available online. URL: http://www.noaanews.noaa.gov/stories/s1069.htm. Posted on December 3, 2002. This article reports on the convening of the U.S. Climate Change Science Program's Planning Workshop for Scientists and Stakeholders for the purpose of reviewing the Strategic Plan Climate Change Science Program. This document sets the direction of climate change research initiatives led by the U.S. government, and directly responds to President Bush's request that the best scientific information be developed to assist the United States in developing a well-reasoned approach to global climate change issues.

OZONE DESTRUCTION AND GLOBAL COOLING

BOOKS

Bryson, Reid A., and Thomas J. Murray. *Climates of Hunger: Mankind and the World's Changing Weather.* Madison: University of Wisconsin Press, 1977. A dip in global temperature between 1950 and 1980 prompted many scientists in the 1970s to predict that climate change was occurring in the form of a profound cooling of the planet. This book, one of the more popular of the era, warns that cooling could disrupt cultural institutions and agricultural practices that have been developed during the past 1,000 years, a time of unusually warm and stable climate. Predictions of imminent drought and famine mirror theories of the effects of global warming, illustrating both the uncertainties scientists face when studying the future of Earth's climate and the power that the idea of climate change, whether warmer or colder, holds on the human imagination.

Cagin, Seth, and Philip Dray. *Between Earth and Sky: How CFCs Changed Our World and Endangered the Ozone Layer.* New York: Pantheon, 1993.

This book documents the history of the use of chlorofluorocarbons CFCs, from their invention in 1928 by General Motors scientist Thomas Midgley, Jr., to their regulation by the 1987 Montreal Protocol, the first global environmental treaty. A narrative history rather than a technical document, *Between Earth and Sky* is written in a journalistic style that is easily accessible to the general reader. It includes notes, a selected bibliography, an index, and black-and-white photographs.

Christie, Maureen. *The Ozone Layer.* Cambridge, England: Cambridge University Press, 2000. Part I covers the history of the study of stratospheric ozone, beginning with its discovery during the 19th century and progressing to research into chlorofluorocarbons and the growing hole over the Antarctic. In Part II, Christie uses ozone as a template to explore the philosophy of scientific inquiry in general, covering such topics as prediction, experimentation, weighing of evidence during efforts to develop viable theories, the use of computers, and the development of scientific consensus. Anyone interested in researching challenging questions about the scientific process itself will find *The Ozone Layer* useful. The book includes notes, a list of references, an index, and graphs.

Gilfond, Henry. *The New Ice Age.* New York: Franklin Watts, 1978. This introductory book is of interest as a historical document from a time when many scientists believed that the slight drop in global temperatures between 1945 and 1975, as well as the unusually cold winter of 1976–77, portended the arrival of a new ice age. Around the time of *The New Ice Age*'s publication, however, global temperatures resumed the upward trend that has been in evidence since. Through easily understood text, graphics, and photographs, Gilfond covers the greenhouse effect, past ice ages, sunspot activity, volcanic eruptions, glaciers, and more.

Halacy, D. S., Jr. *Ice or Fire? Surviving Climatic Change.* New York: Harper and Row, 1978. Halacy looks at the implications of warding off the ice age that many scientists believed was imminent following the frigid winter of 1977–78. The first section covers weather modification and the legal aspects of such intentional manipulations. The second section deals with the technological problems that must be overcome to effectively alter the climate, ice age theories, and the probable effects of an increasingly cold climate on humanity.

Litfin, Karen T. *Ozone Discourses: Science and Politics in Global Environmental Cooperation.* New York: Columbia University Press, 1994. Litfin examines the Montreal Protocol, using the process of drafting and implementing the international ozone treaty to analyze the manner in which scientific knowledge can be used as a political tool.

CHAPTER 8

ORGANIZATIONS AND AGENCIES

This chapter provides contact information for organizations and agencies involved in various aspects of climate change. The following entries include national and international research organizations, government agencies, environmental groups, industry organizations, and public-policy research foundations covering all viewpoints on the subject of global warming.

American Meteorological Society (AMS)
URL: http://www.ametsoc.org/AMS
E-mail: amsinfo@ametsoc.org
Phone: (617) 227-2425
Fax: (617) 742-8718
45 Beacon Street
Boston, MA 02108-3693
Founded in 1919, AMS gathers and disseminates information on the atmospheric, oceanic, and hydrologic sciences. The organization provides a wide range of educational programs and services, sponsors more than 12 conferences each year, and publishes nine journals, including the semimonthly *Journal of the Atmospheric Sciences* and the monthly *Bulletin of the American Meteorological Society*.

American Petroleum Institute (API)
URL: http://www.api.org
Phone: (202) 682-8000
1220 L Street, NW
Washington, DC 20005-4070
API is a member-driven trade association that conducts research and issues policy statements relevant to the oil and natural gas industry. The web site provides statistics on fossil fuel exploration and use, as well as news releases of interest to the petroleum industry.

Asia-Pacific Network for Global Change Research (APN)
URL: http://www.apn.gr.jp
E-mail: info@apn.gr.jp
Phone: (81) 78 230 8017
Fax: (81) 78 230 8018

APN Secretariat
5th Floor
IHD Centre Building
1-5-1 Wakinohama Kaigan Dori
Chuo-ku
Kobe 651-0073
Japan
The APN is an intergovernmental network that fosters global environmental change research in the Asia-Pacific region. The 21 APN member countries include Australia, Bangladesh, Cambodia, China, Fiji, India, Indonesia, Japan, Korea, Laos, Malaysia, Mongolia, Nepal, New Zealand, Pakistan, Philippines, Russia, Sri Lanka, Thailand, the United States, and Vietnam. The network conducts research, collects data, and publishes reports and newsletters.

Aspen Global Change Institute (AGCI)
URL: http://www.gcrio.org/agci-home.html
E-mail: agcimail@agci.org
Phone: (970) 925-7376
Fax: (970) 925-7097
100 East Francis Street
Aspen, CO 81611
AGCI is a nonprofit organization that seeks to improve understanding of global environmental problems by convening interdisciplinary meetings of scientists and by developing education programs for K–12 students. Other services include EarthPulse News, which provides global change updates for educators, and EarthPulse Notes, a database of article summaries on environmental topics.

The Beijer International Institute of Ecological Economics
URL: http://www.beijer.kva.se
E-mail: beijer@beijer.kva.se
Phone: (46) 08 673 9500
Fax: (46) 08 15 24 64
P.O. Box 50005
S-104 05 Stockholm
Sweden
Operating under the auspices of the Royal Swedish Academy of Sciences, the Beijer Institute was founded in 1977 to promote interdisciplinary research, primarily involving ecologists and economists, into the problems that arise from interactions between humans and the environment.

Canada Country Study
URL: http://www.ec.gc.ca/climate/ccs/ccs_e.htm
E-mail: enviroinfo@ec.gc.ca
Phone: (800) 668-6767
Environment Canada Inquiry Centre
Ottawa, Ontario K1A 0H3
Canada
Launched in 1996, this program seeks to assess the impacts of climate change across Canada. The web site provides access to the major published works of the study, including six regional scientific reports and a summary of the 1997 Climate Impacts and Adaptation National Symposium.

Canadian Global Change Program (CGCP)
URL: http://www.globalcentres.org/cgcp/

E-mail: cgcp@uvic.ca
Phone: (250) 472-4337
Fax: (250) 472-4830
University of Victoria
P.O. Box 1700 STN CSC
Victoria, British Columbia V8W 2Y2
Canada

This organization conducts research aimed at promoting action in support f sustained development. The group publishes books, reports, and information bulletins.

Carbon Dioxide Information Analysis Center (CDIAC)
URL: http://cdiac.esd.ornl.gov
E-mail: cdiac@ornl.gov
Phone: (865) 574-3645
Fax: (865) 574-2232
Oak Ridge National Laboratory
Building 2001, MS-050
P.O. Box 2008
Oak Ridge, TN 37831

CDIAC is the U.S. Department of Energy's primary climate-change data and information analysis center. Data holdings include records of atmospheric carbon dioxide concentrations, the role of the terrestrial biosphere and the oceans in the biogeochemical cycles of greenhouse gases, long-term climate trends, the effects of elevated carbon dioxide on vegetation, and the vulnerability of coastal areas to rising sea levels. The CDIAC also includes the World Data Center for Atmospheric Trace Gases. The center publishes papers, theses, the *CDIAC Communications* newsletter, and annual reports.

Cato Institute
URL: http://www.cato.org
Phone: (202) 842-0200
Fax: (202) 842-3490
1000 Massachusetts Avenue, NW
Washington, DC 20001-5403

Founded in 1977, the Cato Institute is a nonprofit public-policy research foundation dedicated to such ideals as limited government and a free market economy. The global warming link on the web site brings up a number of articles that seek to debunk global warming theories and suggest that mitigation strategies will place undue burden on the global economy. The institute hosts conferences and publishes books, monographs, and briefing papers, as well as the *Cato Journal* and *Regulation magazine.*

Center for the Study of Carbon Dioxide and Global Change
URL: http://www.co2science.org/center.htm
E-mail: info@ce2science.org
Phone: (480) 966-3719
Fax: (480) 966-0758
P.O. Box 25697
Tempe, AZ 85285-5697

This organization was founded to disseminate information and commentary on the causes and effects of rising levels of atmospheric carbon dioxide. The opinions presented focus primarily on the benefits of a warmer Earth and the economic costs of programs that aim to limit global warming. The site provides a link to *CO_2 Science*

Magazine, a weekly review of scientific research that supports the editors' point of view.

Climate Institute
URL: http://www.climate.org
E-mail: info@climate.org
Phone: (202) 547-0104
Fax: (202) 547-0111
333½ Pennsylvania Avenue, SE
Washington, DC 20003-1148
This nonprofit organization's stated goal is to use symposia, conferences, roundtable discussions, and special briefings to heighten awareness of climate change and identify practical ways to reduce greenhouse emissions. Topics covered on the web site include sea level rise, extreme weather, air quality, ozone depletion, human health, agriculture, and green energy. Special programs cover climate impacts, energy, international cooperation, environmental refugees, and leadership programs. The institute publishes the quarterly *Climate Alert* newsletter.

Climate Prediction Center (CPC)
URL: http://www.nnic.noaa.gov/cpc
Phone: (301) 763-8000
5200 Auth Road
Camp Springs, MD 20746
This department of the National Oceanic and Atmospheric Administration (NOAA) assesses and forecasts the impacts of short-term climate variability with an emphasis on the risks of extreme weather events. The center maintains a climate glossary, holds annual meetings, and publishes educational materials, climate advisories, assessments, summaries, atlases, and winter stratospheric ozone bulletins.

Climate Solutions
URL: http://www.climatesolutions.org
E-mail: info@climatesolutions.org
Phone: (360) 352-1763
Fax: (360) 943-4977
610 Fourth Avenue East
Olympia, WA 98501-1113
Climate Solutions seeks to stop global warming by implementing practical and profitable solutions to climate change in the Pacific Northwest and British Columbia, Canada, that can be used as models for climate action in other regions. The organization publishes reports on global warming, clean energy, and transportation, and offers such programs as the Energy Outreach Center and the Bicycle Commuter Contest.

Coalition for Environmentally Responsible Economies (CERES)
URL: http://www.ceres.org
E-mail: bakal@ceres.org
Phone: (617) 247-0700
Fax: (617) 267-5400
99 Chauncy Street
Sixth Floor
Boston, MA 02111
CERES is a coalition of environmental, investor, and advocacy

groups that advocate sustainable development. It holds an annual conference in April of each year and publishes reports on particular issues (including global warming), annual reports, corporate environmental and sustainability reports, and historical documents.

Competitive Enterprise Institute (CEI)
URL: http://www.cei.org
E-mail: info@cei.org
Phone: (202) 331-1010
Fax: (202) 331-0640
1001 Connecticut Avenue, NW
Suite 1250
Washington, DC 20036
This nonprofit public-policy organization encourages the removal of government regulation in order to establish a system in which private incentives and property rights would be used to protect the environment. A web search on the Global Warming category under the "Research an Issue and Find a Publication" links brings up numerous articles and press releases on the subject. Publications include *The Environmental Source 2002.*

Earth Island Institute
URL: http://www.earthisland.org
Phone: (415) 788-3666
Fax: (415) 788-7324
300 Broadway, Suite 28
San Francisco, CA 94133
Formed in 1982 by David Brower, Earth Island Institute is dedicated to the conservation, preservation, and restoration of the global environment. The group publishes the quarterly *Earth Island Journal.*

Earth Pledge Foundation
URL: http://www.earthpledge.org
Phone: (212) 725-6611
Fax: (212) 725-6774
122 East 38th Street
New York, NY 10016
This nonprofit organization was originally founded to promote the 1992 Earth Summit. It continues to work toward a sustainable future by developing educational projects and promoting dialogue among leaders in industry, politics, and academia. Publications include the books *Our World in Focus*, *Sustainable Cuisine*, and *Sustainable Architecture.*

Environmental Defense
URL: http://www.environmentaldefense.org/home.cfm
E-mail: members@environmentaldefense.org
Phone: (212) 505-2100
Fax: (212) 505-2375
257 Park Avenue South
New York, NY 10010
Among its other programs, this environmental organization, has worked to eliminate threats posed by climate change by helping to draft the 1990 Clean Air Act and aiding oil companies in developing plans to cut greenhouse emissions. The group publishes newsletters, reports, and an annual report on the state of the environment.

Environment Canada
URL: http://www.ec.gc.ca/climate
E-mail: enviroinfo@ec.gc.ca
Phone: (800) 668-6767
Fax: (819) 953-2225
351 St. Joseph Boulevard
Hull, Quebec K1A 0H3
Canada
Environment Canada coordinates environmental policies and programs for the Canadian government. Its web site includes links to such valuable resources as reports and fact sheets, including the Federal Action Plan on Climate Change and the National Action Program on Climate Change.

European Climate Change Programme (ECCP)
URL: http://europa.eu.int/comm/environment/climat/eccp.htm
E-mail: envinfo@cec.eu.int
Phone: (3) 22 298 1800
Fax (3) 22 298 1899
Commissioner for the Environment
B-1049 Brussels
Belgium
The ECCP was founded by the European Commission in 2000 to develop a strategy by which the nations belonging to the European Union could implement the Kyoto Protocol. The web provides access to the ECCP action plan for implementation of the international treaty.

Friends of the Earth
URL: http://www.foe.org
E-mail: foe@foe.org
Phone: (202) 783-7400
Fax: (202) 783-0444
1025 Vermont Avenue, NW
Third Floor
Washington, DC 20005-6303
This international environmental organization (with affiliates in 63 countries) was founded in 1969 by David Brower "to create a more healthy, just world." Campaigns have included efforts to stop construction of dams, ban international whaling, and reform the World Bank. A search on "climate change" on the web site brings up hundreds of articles. Publications include the *Friends of the Earth Annual Report* and the annual *Green Scissors*, which reveals environmentally harmful government programs that the organization believes should be cut.

The George C. Marshall Institute
URL: http://www.marshall.org
E-mail: info@marshall.org
Phone: (202) 296-9655
Fax: (202) 296-9714
1625 K Street
Suite 1050
Washington, DC 20006
Founded in 1984, this nonprofit research group provides technical advice on scientific issues (including climate change) that have an impact on public policy. Although the institute claims to be unbiased, some readers may perceive its papers and reports on global warming as con-

servative and skeptical. Publications include *Climate Models and the National Assessment*, *A Guide to Global Warming*, *Are Human Activities Causing Global Warming?* and *Climate Forecasting: When Models Are Qualitatively Wrong*.

Global Hydrology and Climate Center (GHCC)
URL: http://www.ghcc.msfc.nasa.gov
Phone: (256) 961-7700
320 Sparkman Drive
Huntsville, AL 35805

A joint venture between NASA and academia, the GHCC studies the global water cycle and assesses its effect on the climate. Among the research topics at the center are infrared remote sensing, global temperatures, atmospheric modeling, atmospheric aerosols, and atmospheric chemistry.

Global Warming International Center (GWIC)
URL: http://www.globalwarming.net
Phone: (630) 910-1551
Fax: (630) 910-1561
22W381 75th Street
Naperville, IL 60565-9245

GWIC disseminates information on global warming science and policy to governments, nongovernmental organizations, and industries in more than 145 countries. The center organizes the annual Global Warming International Conference and Expo, and sponsors projects such as Greenhouse Reduction Benchmark, the Himalayan Reforestation Project, and the Extreme Event Index. Publications include the quarterly newsletter *World Resource Review*.

Goddard Institute for Space Studies (GISS)
URL: http://www.giss.nasa.gov
Phone: (212) 678-5500
Fax: (212) 678-5552
2880 Broadway
New York, NY 10025

GISS works cooperatively with universities and research organizations to study natural and man made changes to the environment, with a particular focus on global climate modeling, Earth observations, climate impacts, and atmospheric chemistry. Among its key research objectives is prediction of atmospheric and climate changes in the twenty first century. The online publications database includes more than 1,000 papers, with an emphasis in those published since 1990.

Greenpeace
URL: http://www.greenpeaceusa.org
Phone: (800) 326-0959
702 H Street, NW
Suite 300
Washington, DC 20021
URL: http://www.greenpeace.org/homepage
E-mail: Supporter.Services@ams.greenpeace.org
Phone: (31) 20 523 6222
Fax: (31) 20 523 6200

Greenpeace International
Keizersgracht 176
1016 DW Amsterdam
The Netherlands
This international direct-action environmental group is concerned with stopping the harmful effects of pollution, nuclear waste, genetic engineering, deforestation, and global warming. The group promotes the use of renewable energy and protests the environmentally destructive practices of oil companies. Publications include climate fact sheets, reports, and briefings, as well as the quarterly *Greenpeace Magazine.*

Hadley Centre for Climate Prediction and Research
URL: http://www.metoffice.gov.uk/research/hadleycentre/index.html
E-mail: hadley@metoffice.com
Met Office
London Road
Bracknell, Berkshire RG12 2PW
United Kingdom
The Hadley Centre is one of the primary sources of climate change research and information in the United Kingdom and the world. Scientists there work to understand the Earth's climate systems and develop climate models that represent them; to monitor global and national climate variability and change; and to attribute recent changes in climate to specific factors. Information on all of these projects is available on the web site.

The Heritage Foundation
URL: http://www.heritage.org
E-mail: info@heritage.org
Phone: (202) 546-4400
Fax: (202) 546-8328
214 Massachusetts Avenue, NE
Washington, DC 20002-4999
This think tank, which supports free enterprise and limited government in environmental matters, publishes studies that question the perceived dangers of global warming and the greenhouse effect. Publications include scholarly journals, reviews, and books.

The Independent Institute
URL: http://www.independent.org
E-mail: info@independent.org
Phone:(510) 632-1366
Fax: (510) 568-6040
100 Swan Way
Oakland, CA 94621-1428
This nonprofit public-policy research organization approaches global warming and climate change from a skeptical point of view, questioning the notion that scientific consensus exists on the issue. The institute publishes *The Independent Review*, a journal, and *The Independent* newsletter (both quarterly), as well as a weekly e-mail newsletter, *The Lighthouse.*

Intergovernmental Panel on Climate Change (IPCC)
URL: http://www.ipcc.ch
E-mail: ipcc_sec@gateway.wmo.ch
Phone: (41) 22 730-8202

Fax: (41) 22 730 8025
c/o World Meteorological Organization
7 bis Avenue de la Paix
C.P. 2300
CH-1211 Geneva 2
Switzerland

Founded in 1988 by the World Meteorological Organization (WMO) and the United Nations Environment Programme (UNEP), the IPCC assesses the scientific, technical, and socioeconomic factors involved in anthropogenic climate change. These assessments, considered by many to be the definitive material on the subject of climate change, are based primarily on peer-reviewed scientific literature. The IPCC has produced special reports, technical papers, methodology guidelines, and supporting material, as well as the often-cited 1990, 1995, and 2001 Assessment Reports. Much of this material may be downloaded from the web site.

International Council for Local Environmental Initiatives (ICLEI)
URL: http://www.iclei.org
E-mail: iclei@iclei.org
Phone: (416) 392-1462
Fax: (416) 392-1478
City Hall, West Tower
16th Floor
100 Queen Street, West
Toronto, Ontario M5H 2N2
Canada

The ICLEI works with local governments around the world to improve the global environment and promote sustainable development through the cumulative effects of local actions. The main web site provides links to international offices, including the U.S. office.

International Institute for Sustainable Development (IISD)
URL: http://www.iisd.org
E-mail: info@iisd.org
Phone: (204) 958-7700
Fax: (204) 958-7710
161 Portage Avenue East
Sixth Floor
Winnipeg, Manitoba R3B 0Y4
Canada

The IISD promotes sustainable development by making policy recommendations on international trade and investment, economic policy, natural resource management, and climate change. Among the institute's publications are *Climate Change, Vulnerable Communities, and Adaptation; Intuit Observations on Climate Change; Agriculture and Climate Change;* and *Incentives for Early Action on Climate Change.* Additional information on global warming and other topics can be found on the IISD's Linkages web page (http://iisd.ca/linkages).

International Society of Tropical Foresters (ISTF)
URL: http://www.cof.orst.edu/org/istf
E-mail: istf@igc.apc.org
Phone: (301) 897-8720
Fax: (301) 897-3690

5400 Grosvenor Lane
Bethesda, MD 20814
The ISTF works on an international scale to promote sustainable, sound methods of managing and harvesting the world's tropical forests. It is the North American and Caribbean facilitator of the global Forest, Trees, and People Program, founded in 1987 to emphasize the need for better cooperation and communication in community forestry. The society publishes the quarterly newsletter *ISTF News.*

Lawrence Livermore National Laboratory
URL: http://www.llnl.gov
E-mail: llnlweb@llnl.gov
Phone: (925) 422-1100
Fax: (925) 422-1370
University of California
7000 East Avenue
Livennore, CA 94550-9234
Lawrence Livermore National Laboratory is a U.S. Department of Energy laboratory operating at the University of California. Founded in 1952 for the development of nuclear weapons, its interests have since expanded to include energy, biomedicine, and environmental science. Scientists in the atmospheric science division work to understand the atmospheric effects of pollutants, develop and test improved atmospheric models, understand the role of fossil fuel emissions in determining greenhouse gas and aerosol concentrations, predict the extent to which stratospheric ozone may change as a result of anthropogenic emissions, quantify the natural variability of the climate system, and understand and quantify interactions between the biosphere and climate. The laboratory publishes reports, abstracts, periodicals, brochures, and newsletters on all aspects of its research.

National Academy of Sciences (NAS)
URL: http://www.nas.edu
500 Fifth Street, NW
Washington, DC 20001
The NAS is a private, nonprofit society of scholars dedicated to conducting research in a wide variety of scientific and engineering fields in a manner that will aid human welfare. The academy is also required to advise the federal government on scientific and technical matters. Among its divisions is the Board on Atmospheric Sciences and Climate.

National Aeronautics and Space Administration (NASA)
URL: http://www.nasa.gov
Phone: (202) 358-0000
300 E Street, SW
Washington, DC 20546-0001
NASA is the U.S. government's primary agency for researching and understanding the Earth and for exploring outer space. Several divisions conduct research and collect data relevant to the study of climate change, including Earth Observatory, Global Hydrology and Climate Center (GHCC), and the Goddard Institute for Space Studies (GISS).

National Center for Atmospheric Research (NCAR)
URL: http://www.ncar.ucar.edu
Phone: (303) 497-1000
1850 Table Mesa Drive
Boulder, CO 80305
NCAR was founded in 1960 to conduct research into climate change, changes in atmospheric composition, Earth-Sun interactions, weather formation and forecasting, and the impacts of these atmospheric phenomena on human societies. The center publishes scientific and technical papers, annual reports, and fact sheets.

National Climate Data Center (NCDC)
URL: http://lwf.ncdc.noaa.gov/oa/ncdc.html
E-mail: ncdc.info@noaa.gov
Phone: (828) 271-4800
Fax: (828) 271-4876
Federal Building
151 Patton Avenue
Asheville, NC 28801-5001
The NCDC develops national and global data sets utilized by government agencies and the private sector to maximize resources provided by the climate and to minimize the risks associated with climate variability and weather extremes. The web site provides access to global climate data.

National Environmental Policy Institute (NEPI)
URL: http://www.nepi.org
E-mail: info@nepi.org
Phone: (202) 857-4784
Fax: (202) 833-5977
1401 K Street, NW
Suite M-103
Washington, DC 20005
NEPI is dedicated to improving environmental policy and management by helping to focus the national environmental debate. Regarding climate change, the institute seeks to maximize the effectiveness of mitigation responses by improving scientific understanding of global warming. Publications include the report *Global Environmental Change: Research Pathways for the Next Decade.*

National Environmental Trust (NET)
URL: http://www.environet.org
E-mail:netinfo@environet.org
1200 Eighteenth Street, NW
Fifth Floor
Washington, DC 20036
NET was founded to provide information about environmental problems through public education programs. Global warming is a primary concern, and the organization's web site lists basic facts about harmful effects arising from climate change, solutions to the problem, and ways that ordinary citizens can take action.

National Geophysical Data Center (NGDC)
URL: http://www.ngdc.noaa.gov/ngdc.html
E-mail:ngdc.info@noaa.gov
Phone: (303) 497-6826
Fax: (303) 497-6513

E/GC 325 Broadway
Boulder, CO 80305-3328
The NGDC provides access to global environmental data from satellites and other sources to promote, protect, and improve the economy, security, environment, and quality of life in the United States.

National Ice Center (NIC)
URL: http://www.natice.noaa.gov
Phone: (301) 457-5303
Federal Building #4
4252 Suitland Road
Washington, DC 20395
The NIC provides analysis and forecasts of global sea ice tailored to meet the requirements of U.S. national interests. The center publishes technical reports, as well as sea ice charts and atlases.

National Institute of Environmental Health Studies
URL: http://www.niehs.nih.gov
Phone: (919) 541-3345
111 Alexander Drive
Research Triangle Park, NC 27709
The NIEHS seeks to reduce the incidence and impact of human illness caused by environmental factors. Among the institute's areas of study are the relationships between climate change and phenomena such as mosquito-borne disease, amphibian deaths, and air pollution–related health problems.

National Oceanic and Atmospheric Administration (NOAA)
URL: http://www.noaa.gov
E-mail: answers@noaa.gov
Phone: (202) 482-6090
Fax: (202) 482-3154
14th Street and Constitution Avenue, NW
Room 6217
Washington, DC 20230
This agency of the U.S. government was formed to observe and study the atmosphere, and to study and conserve natural resources. Among its areas of expertise are oceanography, meteorology, and climatology. Among the NOAA's departments are the Climate Prediction Center (CPC); the National Climate Data Center (NCDC); the National Environmental Satellite, Data, and Information Service (NESDIS); the National Geophysical Data Center (NGDC); and the National Ice Center (NIC).

National Science Foundation (NSF)
URL: http://www.nsf.gov
E-mail: info@nsf.gov
Phone: (703) 292-5111
4201 Wilson Boulevard
Arlington, VA 22230
Founded in 1950, the NSF is an independent agency of the U.S. government whose stated purpose is to promote the progress of science; to advance national health, prosperity, and welfare; and to secure national defense. The Environmental Research and Education division of the NSF oversees the Global Change Research Program, a good source of information on topics such as climate

modeling, ecological rates of change, greenhouse gas dynamics, ozone depletion, and sea level change.

National Snow and Ice Data Center (NSIDC)
URL: http://www.nsidc.org
E-mail: nsidc@nsidc.org
Phone: (303) 492-6199
449 UCB
University of Colorado
Boulder, CO 80309-0449
The NSIDC was established by the National Oceanic and Atmospheric Administration (NOAA) to conduct polar and cryospheric research, and to archive and distribute global snow and ice data. As part of its project, the center monitors snow cover, mountain glacier fluctuations, sea ice extent and fluctuation, changes in ice shelves, and global sea level for signals of climate change.

Natural Resources Defense Council (NRDC)
URL: http://www.nrdc.org
E-mail: nrdcinfo@nrdc.org
Phone: (212) 727-2700
Fax: (212) 727-1773
40 West 20th Street
New York, NY 10011
The NRDC relies on law, science, and member support to protect wildlife and wilderness areas around the world. Concerning global warming, its goals have included raising public awareness, urging the U.S. Senate to ratify the Kyoto Protocol , and promoting renewable energy technologies. Publications include *Kingpins of Carbon: How Fossil Fuel Producers Contribute to Global Warming*, *Feeling the Heat in Florida: Global Warming on the Local Level*, and the quarterly magazine *OnEarth*.

New Hope Environmental Services
URL: http://www.nhes.com
E-mail: info@nhes.com
Phone: (434) 295-7462
Fax: (434) 295-7549
5 Boar's Head Lane
Suite 101
Charlottesville, VA 22903
New Hope Environmental Services is an advocacy science consulting firm that produces research and commentary skeptical of catastrophic, human-induced climate change. The web site provides access to the World Climate Report, a biweekly newsletter that seeks to point out the "weaknesses and outright fallacies" of scientific data that has been collected in support of global warming theory.

President's Council on Environmental Quality (CEQ)
URL: http://www.whitehouse.gov/ceq
Phone: (202) 395-5750
Fax: (202) 456-6546
722 Jackson Place, NW
Washington, DC 20503
The CEQ coordinates federal environmental efforts and works to develop environmental policies and initiatives. A search on "global warming" on the web site brings up links to the text of official White

House press releases and presidential speeches on the topic.

Rainforest Action Network (RAN)
URL: http://www.ran.org
E-mail: rainforest@ran.org
Phone: (415) 398-4404
Fax: (415) 398-2732
221 Pine Street
Suite 500
San Francisco, CA 94104
Since its founding in 1985, RAN has worked to preserve tropical rain forests around the world by addressing problems associated with the logging and importation of tropical lumber, cattle ranching in rain forests, and the human rights of those living in and around rain forests. Publications include the RAN Annual Report.

Rainforest Alliance
URL: http://www.rainforest-alliance.org
E-mail: canopy@ra.org
Phone: (212) 677-1900
Fax: (212) 677-2187
665 Broadway
Suite 500
New York, NY 10012
This environmental organization is concerned with promoting sustainable business practices that will protect ecosystems and the people and wildlife that live within them. Programs include Adopt-a-Rainforest and the Sustainable Agriculture Network. The Rainforest Alliance also publishes a bimonthly newsletter, *The Canopy.*

Rainforest Foundation U.S.
URL: http://www.rainforestfoundation.org
E-mail: rffny@rffny.org
Phone: (212) 431-9098
Fax: (212) 431-9197
270 Lafayette Street
Suite 1107
New York, NY 10012
Founded in 1989 by Trudie Styler and musician Sting, the Rainforest Foundation offers assistance to indigenous rain forest populations seeking to protect their homelands from environmental degradation. The organization publishes newsletters and annual reports, which are available through the web site. Other on-line resources include Rainforest Facts, a list of recommended reading, and links to other organizations that focus on rain forest activism.

Reason Foundation
URL: http://www.reason.org
E-mail: gpassantino@reason.org
Phone: (310) 391-2245
Fax: (310) 391-4395
3414 South Sepulveda Boulevard
Suite 400
Los Angeles, CA 90034-6064
This national public-policy research organization promotes economic and individual liberty, and therefore tends to oppose strict federal regulations aimed at curbing greenhouse emissions. Publications include the monthly *Reason Magazine* and the policy research booklets *New Study Distorts Health Benefits of Greenhouse Gas Reduction*, *Q&A About Forests and Global Climate Change*, *Plain*

English Guide 3: Exploring the Science of Climate Change, and *Evaluating the Kyoto Approach to Climate Change.*

Renew America
URL: http://sol.crest.org/environment/renew_america
E-mail: renewamerica@counterpart.org
Phone: (202) 721-1545
Fax: (202) 467-5780
1200 18th Street, NW
Suite 1100
Washington, DC 20036
This nonprofit organization, founded in 1989, promotes the exchange of ideas among community and environmental groups, businesses, politicians, and civic activists for the purpose of improving the environment. Each year, Renew America presents its National Awards for Sustainability to outstanding environmental programs.

Resources for the Future (RFF)
URL: http://www.rff.org
Phone: (202) 328-5000
Fax: (202) 939-3460
1616 P Street, NW
Washington, DC 20036-1400
RFF is a nonprofit think tank that conducts independent, economics-based research on environmental and natural resource issues. Climate change is a major concern of the organization, which publishes numerous project summaries, issue briefs, articles, reports, books, and discussion papers on the subject.

Science and Environmental Policy Project (SEPP)
URL: http://www.sepp.org
E-mail: comments@sepp.org
Phone: (703) 920-2744
1600 South Eads Street
Suite 712-S
Arlington, VA 22202-2907
SEPP is a policy and research organization founded in 1990 by global warming skeptic S. Fred Singer. Research areas include climate change, sea level rise, and El Niño. The web site includes a link to the 1997 Leipzig Declaration, drafted during the International Symposium on the Greenhouse Controversy in opposition to the Kyoto Protocol. SEPP publishes the briefing report *The Scientific Case Against the Global Climate Treaty.*

Sierra Club
URL: http://www.sierraclub.org
E-mail: information@sierraclub.org
Phone: (415) 977-5500
Fax: (415) 977-5799
85 Second Street
Second Floor
San Francisco, CA 94105-3441
The Sierra Club promotes the protection and preservation of natural resources. The group's global warming action network provides information on clean energy alternatives and the clean car campaign. Publications include global warming books and fact sheets, emissions maps, the bimonthly magazine *Sierra,* and the monthly activist resource *The Planet.*

Stockholm Environment Institute (SEI)
URL: http://www.seib.org
E-mail: seib@tellus.org
Phone: (617) 266-8090
Fax: (617) 266-8303
11 Arlington Street
Boston, MA 02116-3411
Headquartered in Sweden, SEI is an international research institute that focuses on sustainable development and a variety of environmental issues, including climate change and energy use. The U.S. office, based in Boston, is part of the Tellus Institute (www.tellus.org), which promotes environmental stewardship and equitable development.

Union of Concerned Scientists (UCS)
URL: http://www.ucsusa.org/index.html
E-mail: menu@ucsusa.org
Phone: (617) 547-5552
Fax: (617) 864-9405
2 Brattle Square
Cambridge, MA 02238-9105
This alliance of 50,000 citizens and scientists works to reduce the misuse of science and technology in society by redirecting research toward what they see as pressing environmental and social problems. These programs include the impacts of global warming, renewable energy technologies, transportation reform, and sustainable agriculture. Publications include the reports *Common Sense on Climate Change: Practical Solutions to Global Warming*, *Global Warming: Early Warning Signs*, and *A Small Price to Pay: U.S. Action to Curb Global Warming Is Feasible and Affordable.*

University Corporation for Atmospheric Research (UCAR)
URL: http://www.ucar.edu
Phone: (303) 497-1000
1850 Table Mesa Drive
Boulder, CO 80305
UCAR was founded in 1959 to enhance the ability of universities to observe and collect data about the atmosphere. The organization supplies up-to-date weather data to universities, trains weather forecasters, and helps organize international research efforts. Through its National Center for Atmospheric Research (NCAR), UCAR also maintains state-of-the-art research technologies, such as computer models, that are used by scientists around the world.

United Nations Environment Programme (UNEP)
URL:http://www.unep.org
E-mail: eisinfo@unep.org
Phone: (254-2) 621234
Fax: (254-2) 624489/90
United Nations Avenue, Gigiri
P.O. Box 30552
Nairobi, Kenya
UNEP was established in 1972 to encourage sustainable development through sound environmental practices on a global scale. Among its areas of focus are the atmosphere and climate change. Its publications include annual re-

ports and the Global Environmental Outlook series.

United Nations International Strategy for Disaster Reduction (UNISDR)
URL: http://www.unisdr.org
E-mail: isdr@un.org
Phone: (41) 22 917 9711
Fax: (41) 22 91790 98/99
Palais des Nations
CH-1211 Geneva 10
Switzerland
The UNISDR is concerned with mitigating the effects of natural disasters on human populations, including those disasters that many scientists believe will be amplified by global warming: Floods, drought, and El Niño, for example.

U.S. Energy Information Administration (EIA)
URL: http://www.eia.doe.gov
E-mail: infoctr@eia.doe.gov
Phone: (202) 586-8800
EI 30
1000 Independence Avenue, SW
Washington, DC 20585
The EIA, created by the U.S. Congress in 1977, is the statistical agency of the U.S. Department of Energy. It provides policy-independent data, forecasts, and analysis to promote sound policy making, efficient markets, and public understanding regarding energy and its interaction with the economy and the environment. The EIA publishes pamphlets, directories, and feature articles on a variety of energy-related subjects, while the web site provides access to current energy data, a glossary, and other information.

U.S. Environmental Protection Agency (EPA)
URL: http://www.epa.gov
E-mail: public-access@epa.gov
Phone: (202) 260-2090
Ariel Rios Building
1200 Pennsylvania Avenue, NW
Mail Code 3213A
Washington, DC 20460
The mission of the EPA is to protect human health and to safeguard the natural environment upon which life depends. The EPA's Office of Air and Radiation maintains statistics on greenhouse gas emissions and global climate change. The office publishes sea level rise reports, greenhouse gas emissions reports, documents providing background on the U.S. position concerning global warming, EPA conference reports, and other documents.

U.S. Global Change Research Information Office (GCRIO)
URL: http://www.gcrio.org
E-mail: help@gcrio.org
Phone: (845) 365-8930
Fax: (845) 365-8922
P.O. Box 1000
61 Route 9W
Palisades, NY 10964
GCRIO provides access to data on global change research, adaptation and mitigation strategies and technologies, and education resources on behalf of U.S. federal agencies involved in the U.S. Global Change Research Program. The office makes

available key documents and reports generated or sponsored by the U.S. government, including the National Assessment of the Potential Consequences of Climate Variability and Change, conducted as part of the major ongoing effort to understand what climate change means for the United States.

U.S. Global Climate Change Research Program (USGCRP)
URL: http://www.usgcrp.gov
E-mail: information@usgcrp.gov
Phone: (202) 488-8630
Fax: (202) 488-8681
400 Virginia Avenue
Suite 750
Washington, DC 20024
This national research program was created to address uncertainties about natural and human-induced global environmental change; monitor, understand, and predict global change; and provide a sound scientific basis for national and international decision making. Among its publications are national and international assessments of climate change and ozone depletion.

U.S. Office of Energy Efficiency and Renewable Energy (EERE)
URL: http://www.eren.doe.gov
Phone: (303) 275-4700
Fax: (303) 275-4788
Golden Field Office
1617 Cole Boulevard
Golden, CO 80401-4700
Part of the U.S. Department of Energy, the EERE was formed to revolutionize approaches to energy efficiency and renewable energy technologies. Research areas include geothermal, hydrogen, ocean, solar, and wind power.

U.S. Office of Scientific and Technical Information (OSTI)
URL: http://www.osti.gov
E-mail: ostiwebmaster@osti.gov
Phone: (865) 576-1188
Fax: (865) 576-2865
P.O. Box 62
Oak Ridge, TN 37831
OSTI is responsible for disseminating scientific and technical information resulting from U.S. Department of Energy research and development programs. OSTI has produced EnergyFiles (a vast online library of information and resources pertaining to energy science and technology), the DOE Information Bridge (providing access to full-text reports), and PubScience (providing access to scientific literature).

Woods Hole Research Center (WHRC)
URL: http://www.whrc.org
E-mail: info@whrc.org
Phone: (508) 540-9900
Fax: (508) 540-9700
P.O. Box 296
Woods Hole, MA 02543-0296
The WHRC addresses environmental issues through scientific research and education and through the application of science to public affairs. The center focuses on global warming, particularly the controlling influence of global forests on climate. Publications include essays, ab-

stracts, and the book *You Can't Eat GNP: Economics as If Ecology Mattered.*

World Bank
URL: http://www.worldbank.org
Phone: (202) 473-1000
Fax: (202) 477-6391
1818 H Street, NW
Washington, DC 20433
The World Bank provides loans to developing economies for the purpose of improving living standards and eliminating the worst forms of poverty. As global warming is expected to have a disproportionate effect on developing countries, the World Bank has increasingly taken potential problems arising from climate change into account. Strategies include promoting energy efficiency, conservation, and renewable energy; reducing long-term costs of low-emission energy technologies; and supporting the development of sustainable transportation. The World Bank publishes several reports dealing with greenhouse gas analysis, energy strategies, and mitigation.

World Conservation Union
URL: http://www.iucn.org
E-mail: mail@hq.iucn.org
Phone: (41) 22 999 0000
Fax: (41) 22 999 0002
Rue Mauverney 28
Gland 1196
Switzerland
Founded in 1948, this organization works in 140 countries to encourage the conservation of nature and to ensure that natural resources are used in an equitable and ecologically sustainable manner. Among its programs is the Climate Change Initiative, formed to develop strategies to confront global warming and to assist nations in climate change adaptation and mitigation. Publications include *Climate, Biodiversity, and Forests: Issues and Opportunities Emerging from the Kyoto Protocol* and *Environmental Law Reporter: Can the Kyoto Protocol Support Biodiversity Conservation?*

World Data Center for Paleoclimatology
URL: http://www.ngdc.noaa.gov/paleo/paleo.html
E-mail:paleo@noaa.gov
Phone: (303) 497-6280
Fax: (303) 497-6513
325 Broadway
Code E/GC
Boulder, CO 80305-3328
The Paleoclimatology Program was set up by the National Oceanic and Atmospheric Administration (NOAA) to study past climates in order to better predict future climate change. Links on the web site lead to explanations of how paleoclimate research relates to drought and global warming.

World Energy Council (WEC)
URL: http://www.worldenergy.org/wec-geis
E-mail: info@worldenergy.org
Phone: (44) 207 734 5996
Fax: (44) 207 734 5926
1-4 Warwick Street
Regency House
5th Floor
London W1B 5LT
United Kingdom

With member committees in more than 90 countries, the United Nations–accredited WEC is the foremost multi-energy organization in the world. Its goal is to promote sustainable supply and use of energy on a global scale. Among its publications are an annual report, the *Survey of Energy Resources*, and numerous papers and studies.

World Meteorological Organization (WMO)
URL: http://www.wmo.ch
E-mail: ipa@www.wmo.ch
Phone: (41) 22 730 8111
Fax: (41) 22 730 8181
7 bis Avenue de la Paix
CP 2300
1211 Geneva 2
Switzerland
The United Nations–affiliated WMO coordinates global scientific activity to develop increasingly prompt and accurate weather and climate information for public, private, and commercial use.

World Resources Institute (WRI)
URL: http://www.wri.org
E-mail: front@wri.org
Phone: (202) 729-7600
Fax: (202) 729-7610
10 G Street, NE
Suite 800
Washington, DC 20002
WRI is a think tank that provides information on, ideas about, and solutions to global environmental problems. It conducts research into global resources and environmental issues, sponsors seminars and conferences, and provides media outlets with background materials on environmental issues. Publications include books, reports, and papers, many on global climate change.

Worldwatch Institute
URL: http://www.worldwatch.org
E-mail: worldwatch@worldwatch.org
Phone: (202) 452-1999
Fax: (202) 296-7365
1776 Massachusetts Avenue, NW
Washington, DC 20036-1904
This nonprofit public-policy research organization works to focus the attention of policy makers and the public on global problems, including environmental concerns such as global warming and the relationship between the world economy and the environment. The institute publishes the reports *Vital Signs* and *State of the World* (both annual), the bimonthly magazine *World Watch*, and numerous papers, books, and CD-ROMs.

World Wildlife Fund (WWF)
URL: http://www.panda.org
Phone: (202) 293-4800
Fax: (202) 293-9211
1250 24th Street, NW
Washington, DC 20037-1175
This global environmental organization works to conserve biological diversity, ensure that natural resources are managed sustainably, and reduce pollution and consumption. Among its priorities is reducing greenhouse gas emissions in order to lessen the effects of global warming.

PART III

APPENDICES

APPENDIX A

GLOBAL WARMING TRENDS AND THE GREENHOUSE EFFECT

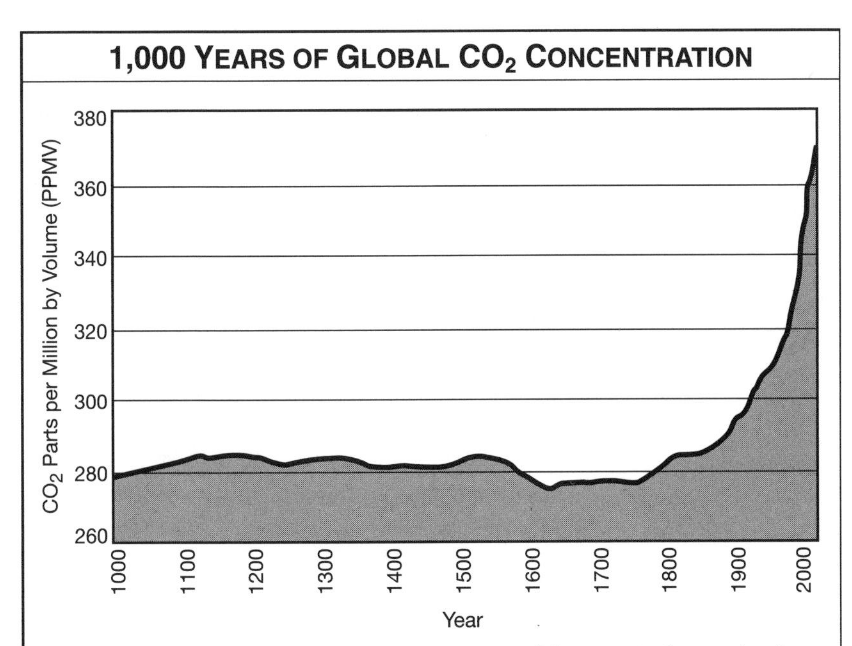

Records of Northern Hemisphere surface temperatures, CO_2 concentrations, and carbon emissions show a close correlation. This chart shows a record of global CO_2 concentration for 1,000 years derived from measurements of CO_2 concentration in air bubbles in the layered ice cores drilled in Antarctica and from atmospheric measurements since 1957.

Source: National Assessment Synthesis Team. *Climate Change Impacts on the United States: The Potential Consequences of Climate Variability and Change*. Washington, D.C.: U.S. Global Change Research Program, 2000.

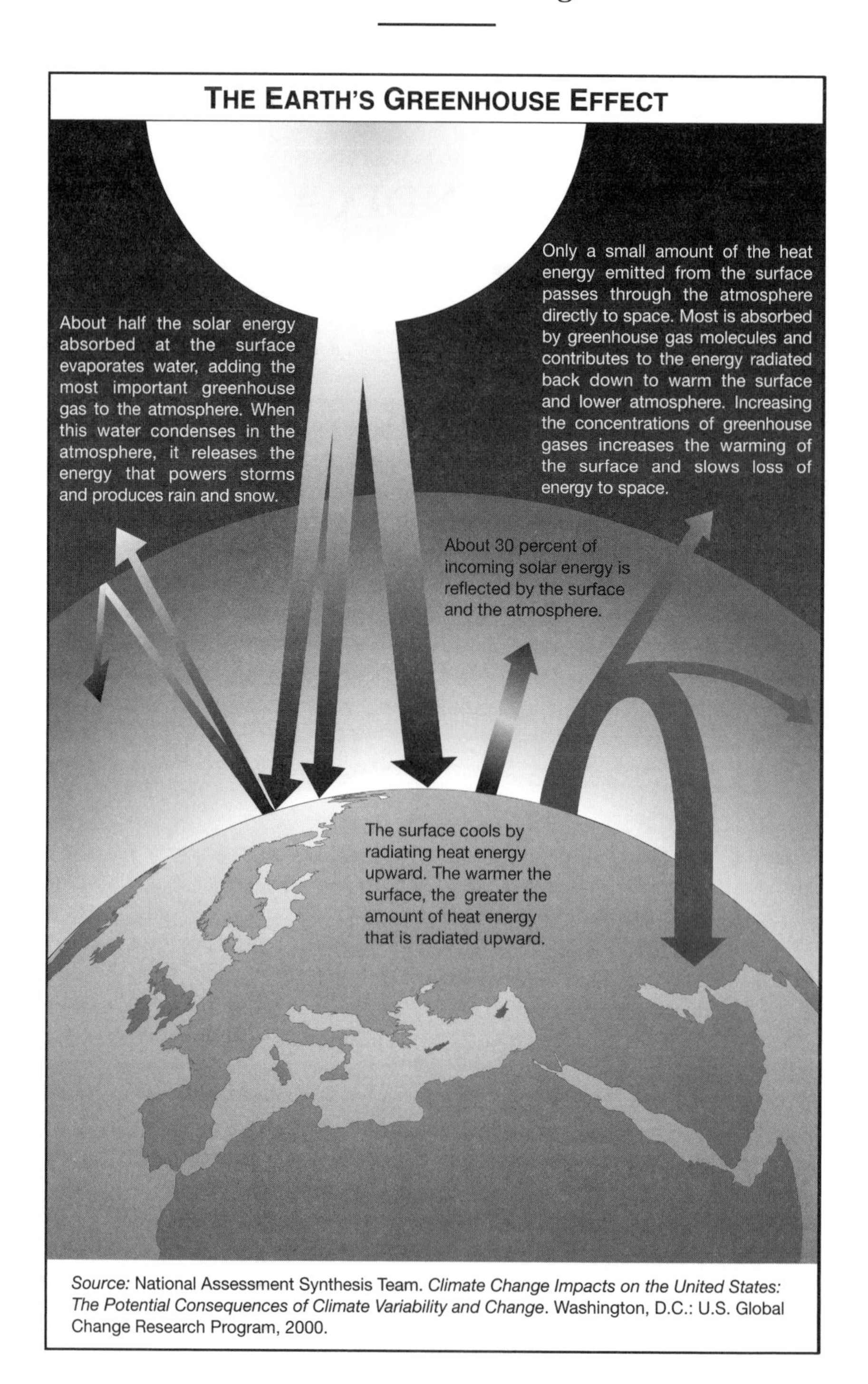

Source: National Assessment Synthesis Team. *Climate Change Impacts on the United States: The Potential Consequences of Climate Variability and Change*. Washington, D.C.: U.S. Global Change Research Program, 2000.

APPENDIX B

EXCERPTS FROM THE ENERGY POLICY ACT OF 1992, H.R. 776, PUBLIC LAW: 102-486 (10/24/1992)

Title I: Energy Efficiency—Subtitle A: Buildings—Amends the Energy Conservation and Production Act to set a deadline by which each State must certify to the Secretary of Energy (the Secretary) whether its energy efficiency standards with respect to residential and commercial building codes meet or exceed those of the Council of American Building Officials (CABO) Model Energy Code, 1992, and of the American Society of Heating, Refrigerating, and Air-Conditioning Engineers, respectively.

Requires the Secretary to provide technical assistance and incentive funding to the States to promote increased use of energy efficiency codes for buildings. Authorizes appropriations.

Directs the Secretary to: (1) establish standards that require energy efficiency measures that are technologically feasible and economically justified in new Federal buildings; and (2) review them every five years. Mandates Federal agency compliance with such standards. Prescribes guidelines under which the Secretary shall support the upgrading of voluntary building energy codes for new residential and commercial buildings.

Amends the Cranston-Gonzalez National Affordable Housing Act to require the Secretary of Housing and Urban Development (HUD), and the Secretary of Agriculture to jointly establish energy efficiency standards for residential housing. (Currently only the Secretary of HUD is required to do so). Amends Federal law regarding veterans' readjustment benefits to condition a loan for new residential housing upon compliance with such standards.

Amends the National Energy Conservation Policy Act to direct the Secretary of Energy to: (1) issue voluntary building energy code guidelines for use by the private and public sectors to encourage the assignment of energy efficiency ratings for new residential buildings; (2) establish a technical assistance program for State and local organizations to encourage the use of residential energy efficiency rating systems consistent with such guidelines; (3) provide matching grants for the establishment of regional building energy efficiency centers in each of the regions served by a Department of Energy (DOE) regional support office; and (4) establish an advisory task force to evaluate grant activities. Authorizes appropriations.

Requires the Secretary of HUD to: (1) assess the energy performance of manufactured housing and make recommendations to the National Commission on Manufactured Housing regarding thermal insulation and energy efficiency improvements; and (2) test the performance and determine the cost effectiveness of manufactured housing constructed in compliance with certain statutory standards. Authorizes the States to establish thermal insulation and energy efficiency standards for manufactured housing if the Secretary of HUD has not issued final regulations by October 1993.

Amends the Cranston-Gonzalez National Affordable Housing Act to direct the Secretary of HUD to promulgate a uniform affordable housing plan using energy efficient mortgages (mortgages that provide financing incentives either for the purchase of energy efficient homes, or for incorporating the cost of such improvements into the mortgage).

Requires the Secretary of HUD to establish an energy efficient mortgage pilot program in five States under prescribed guidelines in order to promote the purchase of existing energy efficient residential buildings and the installation of cost-effective improvements in existing residential buildings. Authorizes appropriations.

Subtitle B: Utilities—Amends the Public Utility Regulatory Policies Act of 1978 (PURPA) to mandate that: (1) each electric utility employ integrated resource planning; (2) the rates for a State regulated electric utility are such that its outlay for demand side management measures (including energy conservation and energy efficiency resources), are at least as profitable as those for the construction of new generation, transmission, and distribution equipment; (3) the rates charged by any electric utility are such that it is encouraged to make outlays for all cost-effective improvements in energy efficient power generation, transmission, and distribution; and (4) such rates and charges are implemented in a manner that assures that utilities are not granted unfair competitive advantages over small businesses engaged in transactions regarding demand side energy management measures.

Prescribes guidelines under which the Secretary of Energy may provide grants to State regulatory authorities to encourage demand-side management (including energy conservation, efficiency, and load management techniques), and as a means of meeting gas supply needs. Permits a State regulatory authority to provide financial assistance to nonprofit subgrantees of the DOE Weatherization Assistance Program. Authorizes appropriations.

Directs the Tennessee Valley Authority (TVA) to conduct a least-cost planning program according to prescribed guidelines. Exempts TVA from specified least-cost planning requirements which might arise from its electric power transactions with the Southeastern Power Administration.

Amends the Hoover Power Plant Act of 1984 to establish guidelines within which the Administrator of the Western Area Power Administration shall require each customer purchasing electric energy under a long-term firm power service contract with the Western Area Power Administration to implement integrated resource planning by October 1995. Permits different regulations for certain small customers that have limited capability to conduct integrated resource planning. Authorizes the Administrator to provide customers with technical assistance to implement such resource planning. Sets forth: (1) a schedule within which each customer must submit an initial integrated resource plan and periodic revisions to the Administrator; and (2) approval criteria for integrated resource plans. Provides for enforcement of integrated resource plan requirements, including the imposition of a surcharge, and a reduction in the power allocation of a noncomplying customer. Permits customers to form integrated resource planning cooperatives.

Amends PURPA to mandate that: (1) gas utilities employ integrated resource planning for gas customers; and (2) the rates charged by a State regulated gas utility are such that its prudent outlays for demandside measures (such as energy conservation and load shifting programs) are at least as profitable as its outlays for supplies and facilities. Requires a State regulatory authority to implement its integrated resource planning standards in a manner that assures that utilities are not provided with unfair competitive advantages over small businesses.

Subtitle C: Appliance and Equipment Energy Efficiency Standards—Directs the Secretary to provide financial assistance to support a voluntary national window rating program that will develop energy ratings and labels for windows and window systems. Requires the National Fenestration Rating Council to develop such rating program according to specified procedures. Requires the Secretary to develop specified alternative rating systems if a national voluntary window rating program consistent with this Act has not been developed.

Amends the Energy Policy and Conservation Act to: (1) detail energy conservation and labeling requirements for specified commercial and industrial equipment (including lamps and plumbing products); and (2) delineate standards for heating and air-conditioning equipment, electric motors, high intensity discharge lamps, and distribution transformers.

Directs the Secretary to provide financial and technical assistance to support a voluntary national testing and information program for widely used commercial office equipment and luminaries with potential for significant energy savings.

Requires the Secretary to report to the Congress on: (1) the potential for the development and commercialization of appliances which are substantially more efficient than required by Federal or State law; and (2) the energy savings and environmental benefits of early appliance replacement programs.

Subtitle D: Industrial—Directs the Secretary to: (1) make matching grants to industry associations for energy efficiency improvement programs; (2) establish an annual recognition award program for industrial entities that have significantly improved their energy efficiency; and (3) report to the Congress on the establishment of Federally mandated energy efficiency reporting requirements and voluntary energy efficiency improvement targets for energy intensive industries.

Directs the Secretary to make renewable grants to the States to: (1) promote the use of energy-efficient technologies in covered industries; (2) establish industry-by-industry training programs to conduct process-oriented industrial assessments; (3) assist utilities in developing energy efficiency programs and technologies for industrial customers in covered industries; (4) establish and update criteria for conducting process-oriented industrial assessments on an industry-by-industry basis; (5) establish a nationwide directory of organizations offering industrial energy efficiency assessments, technologies, and services; (6) establish an annual recognition award program for utilities operating outstanding or innovative industrial energy efficiency technology assistance programs; and (7) convene annual meetings of parties interested in process-oriented industries. Authorizes appropriations.

Requires the Secretary to: (1) establish voluntary guidelines for energy efficiency audits and insulation in industrial facilities; and (2) conduct educational and technical assistance programs to promote their use.

Subtitle E: State and Local Assistance—Amends the Energy Policy and Conservation Act to: (1) authorize the Secretary to provide funds to specified States to finance energy efficiency improvements in State and local government buildings; and (2) expand the optional features of State energy conservation programs.

Amends the Energy Conservation and Production Act to: (1) direct the Secretary to provide financial assistance to governmental weatherization assistance recipients for the development and initial implementation of private sector arrangements under which non-Federal financial assistance would be made available to support energy efficiency improvement programs for low-income housing; and (2) authorize the Secretary to provide financial assistance to such recipients for energy efficiency technical transfer measures.

Repeals the National Energy Extension Service Act.

Subtitle F: Federal Agency Energy Management—Amends the National Energy Conservation Policy Act to set a deadline by which each Federal agency must install energy and water conservation measures with payback periods of less than ten years. Requires the President to transmit in the annual budget request to the Congress a statement of the amount of appropriations requested on an individual agency basis for energy conservation measures in Federal facilities.

Requires the Secretary to: (1) develop a mechanism for Federal agencies to implement an energy conservation incentive program; and (2) establish a Federal Energy Efficiency Fund to provide grants to agencies to meet Federal energy management requirements.

Authorizes Federal agencies to participate in utility incentive programs for energy and water conservation. Directs the Secretary to: (1) establish a financial bonus program for outstanding Federal facility energy managers; (2) establish an energy conservation technology demonstration program for installation in Federal facilities or federally assisted housing; and (3) implement a survey of potential energy savings in Federal buildings.

Amends the Federal Property and Administrative Services Act of 1949 to cite conditions under which the Administrator of General Services may obligate: (1) funds from the Federal Building Fund for energy management improvement and source reduction and recycling programs; and (2) goods and services received from a utility which enhance the energy efficiency of Federal facilities.

Amends the National Energy Conservation Policy Act to prescribe the parameters of statutorily mandated energy savings performance contracts entered into by a Federal agency.

Requires the Administrator of General Services to hold regular, biennial conference workshops in each of the ten standard Federal regions on energy management, conservation, efficiency and planning strategy.

Requires specified Federal executive departments to undertake programs to ensure full training of facility energy managers.

Directs the Secretary of Energy to make energy audit teams available to all Federal agencies. Requires the Director of the Office of Management and Budget to establish guidelines for Federal agencies to use in assessing energy consumption in their facilities.

Sets a deadline by which certain Inspectors General must: (1) identify agency compliance activities to meet specified requirements of the National Energy Conservation Policy Act; and (2) determine if such agencies have the requisite internal accounting mechanisms to assess the accuracy and reliability of energy consumption and energy cost figures pursuant to such Act. Requires the President's Council on Integrity and Efficiency to report to certain congressional committees regarding such Inspector General reviews.

Prescribes guidelines under which the Administrator of General Services, the Secretary of Defense, and the Director of the Defense Logistics Agency must include energy efficient products in their procurement and supply operations.

Directs the Secretary of Energy to study and report to the Congress on financing options for certain statutorily mandated energy and water conservation measures.

Requires the Postmaster General to: (1) conduct a prescribed energy survey for Postal Service buildings; (2) report to certain congressional committees on the building management program as it related to energy efficiency; and (3) ensure that each Postal Service facility meets certain statutorily mandated energy management requirements.

Requires each agency to establish criteria for the improvement of energy efficiency in Federal facilities operated by Federal Government contractors.

Directs the Architect of the Capitol to retrofit congressional buildings according to prescribed energy conservation guidelines. Authorizes appropriations.

Subtitle G: Miscellaneous—Amends the Department of Energy Organization Act to direct the Administrator of the Federal Energy Administration to engage in energy information data collection including: (1) electricity production from domestic renewable energy resources; (2) residential and commercial energy; and (3) demand-side management programs conducted by electric utilities.

Requires the Secretary to study and report to the Congress on specified aspects of district heating and cooling programs and vibration reduction technologies.

Title III: Alternative Fuels—General—Amends the Energy Policy Conservation Act to prescribe guidelines under which the Secretary of Energy shall: (1) acquire alternative fueled vehicles for the Federal fleet; and (2) study and report to the Congress on the Federal experience with alternative-fueled heavy duty vehicles. Sets forth a fiscal-year schedule for minimum Federal fleet requirements for alternative-fueled vehicles. Directs the Secretary to provide guidance and technical assistance to Federal agencies for the procurement and placement of alternative-fueled vehicles.

Directs the Administrator of General Services (the Administrator) to establish an annual recognition and incentive awards program for Federal employees who demonstrate the strongest commitment to the use of alternative fuels and fuel conservative in Federal motor vehicles. Authorizes appropriations.

Directs the Administrator and the Postmaster General to report to the Congress on the alternative-fueled vehicle program within their respective agencies.

Title IV: Alternative Fuels—Non-Federal Programs—Amends the Energy Policy and Conservation Act to authorize appropriations for FY 1993 through 1995 for the alternative fuels truck commercial application program.

Amends the Motor Vehicle Information and Cost Savings Act to reflect the provisions of this Act regarding the use of alternative motor fuels.

Amends the Natural Gas Act to exempt from its jurisdiction certain vehicular natural gas sales and transportation transactions if the person involved is: (1) not otherwise a natural-gas company; and (2) subject to a State regulatory agency.

Cites circumstances under which certain holding companies engaged in vehicular natural gas transactions are exempt from the purview of Public Utility Holding Company Act of 1956.

Directs the Secretary of Energy (the Secretary) to establish a public information program on the use of alternative fuels in motor vehicles. Requires the Federal Trade Commission to formulate and issue rules for labeling requirements for alternative fuels and alternative fueled vehicles. Requires the Secretary to establish a data collection program for persons engaged in certain activities related to alternative-fuel vehicles and facilities.

Authorizes FERC to allow recovery of expenses in advance by natural-gas companies and electric utilities, respectively, for certain research and demonstration activities relating to natural-gas motor vehicles and electric motor vehicles.

Directs the Secretary to promulgate guidelines for comprehensive State program plans and incentives to accelerate the introduction and use of alternative fuels and alternative-fueled vehicles.

Authorizes the Secretary of Transportation to enter into cooperative agreements and joint ventures with governmental and regional transit authorities of certain-sized urban areas to demonstrate the feasibility of commercial application of alternative fuels for motor vehicles used for mass transit (including school buses).

Directs the Secretary to: (1) ensure that a Federal certification program is established for technician training programs for converting conventionally-fueled motor vehicles into dedicated or dual-fueled vehicles; (2) conduct a study to determine whether the use of alternative fuels in nonroad vehicles and engines would contribute substantially to reduced reliance on imported energy sources; (3) report to the Congress on how Federal purchasing policies and governmental traffic control policies affect the use of alternative-fueled vehicles; and (4) establish a low-interest loan program to increase the use by small businesses of alternative-fueled vehicles. Authorizes appropriations.

Title V: Availability and Use of Replacement Fuels, Alternative Fuels, And Alternative Fueled Private Vehicles—Sets forth a timetable for the acquisition of alternative fueled vehicles by specified persons engaged in fuels transactions.

Requires the Secretary to: (1) establish a program to promote the development and use in light duty motor vehicles of domestic replacement fuels in lieu of petroleum motor fuels; (2) estimate annually the use and supply of alternative and replacement fuels, and their affect upon greenhouse gas emissions; (3) require suppliers of replacement and alternative fuels to submit specified supply and greenhouse gas–related information to the Secretary; and (4) undertake to obtain voluntary commitments from specified persons to make available to the public replacement fuels and alternative fueled vehicles (and attendant services).

Sets forth specified fleet program purchase goals for alternative-fueled vehicles for specified calendar years, (including mandatory State fleet programs). Sets forth civil penalties for violations of this Act. Authorizes appropriations.

Title VI: Electric Motor Vehicles—Subtitle A: Electric Motor Vehicle Commercial Demonstration Program—Directs the Secretary to conduct an electric motor vehicle demonstration program. Details the program's proposal parameters. Authorizes appropriations.

Subtitle B: Electric Motor Vehicle Infrastructure and Support Systems Development Program—Directs the Secretary to undertake, with non-Federal persons, an electric motor vehicle infrastructure and support systems development program. Sets forth program parameters. Authorizes appropriations.

Title XII: Renewable Energy—Amends the Renewable Energy and Efficiency Technology Competitiveness Act of 1989 to direct the Secretary to: (1) implement a five-year program to further the commercialization of renewable energy and energy efficiency technologies by soliciting proposals for demonstration and commercial application projects; and (2) establish an Advisory Committee on Demonstration and Commercial Application of Renewable Energy and Energy Efficiency Technologies. Authorizes appropriations.

States that the goal of the Alcohol from Biomass Program is to advance research and development to a point where alcohol from biomass technology is cost-competitive with conventional hydrocarbon transportation fuels, and to promote the integration of such technology into the transportation fuel sector of the economy. Delineates goals for producing ethanol from biomass.

Directs the Secretary to: (1) prepare and submit to the Congress a three-year national renewable energy and energy efficiency management plan with specified contents; (2) establish a renewable energy export technology training program for individuals from developing countries; (3) make Renewable Energy Advancement Awards in recognition of developments that advance the practical application of certain renewable energy technologies; and (4) study and report to the Congress on whether certain conventional taxation and ratemaking procedures result in economic barriers to, or incentives for, renewable energy power plants compared to conventional power plants. Authorizes appropriations.

Directs the Department of Energy to conduct a study to facilitate the marketing of energy byproducts from rice milling.

Amends the Energy Policy and Conservation Act to: (1) direct the interagency working group to make recommendations to coordinate Federal actions and programs affecting reports of renewable energy and energy efficiency products and services; and (2) establish an Interagency Working Subgroup on Renewable Energy and an Interagency Working Subgroup on Energy Efficiency to recommend coordinated Federal actions and programs to promote the export of domestic renewable energy and energy efficiency products and services, and to promote their development and application in foreign countries. Authorizes appropriations.

Directs the interagency working group to study and report to the Congress on the export promotion practices foreign countries use with respect to their own renewable energy and energy efficiency technologies and products.

Directs the Secretary of Commerce to develop, and make available to interested persons, a comprehensive data base and information dissemination system on the specific energy technology needs of foreign countries, and the technical and economic competitiveness of various renewable energy and energy efficiency products and technologies; and (2) select individuals experienced in renewable energy and energy efficiency products and technologies to be assigned to an office of the United States and Foreign Commercial Service in the Pacific Rim, and in the Caribbean Basin, respectively, in order to provide information concerning domestic renewable energy and energy efficiency products, technologies, and industries. Authorizes appropriations.

Requires the Secretary of Energy to: (1) establish a renewable energy technology transfer program to implement itemized purposes; and (2) make incentive payments to the owner or operator of a qualified renewable energy facility for a 20-year period for any electricity generated and sold. Authorizes appropriations.

Title XVI: Global Climate Change—Requires the Secretary of Energy to report to the Congress on specified implications of global climate change policies, including the generation of greenhouse gases and carbon dioxide, and U.S. compliance with its international obligations.

Requires each National Energy Policy Plan submitted by the President to include a least-cost energy strategy prepared by the Secretary.

Directs the Secretary to establish a Director of Climate Protection within DOE to serve as the Secretary's representative for interagency and multilateral policy discussion of global climate change, and to monitor the effects of domestic and international policies upon greenhouse gas generation.

Requires the Secretary to: (1) present a comparative assessment report to the Congress on alternative policy mechanisms for reducing greenhouse gas generation; (2) develop an inventory of the national aggregate emissions of each greenhouse gas for a specified baseline period; (3) issue guidelines for the voluntary collection and reporting of information on sources of greenhouse gases; and (4) establish a data base composed of such information.

Amends the Energy Security Act to repeal the mandate that the President transmit specified biennial energy targets to the Congress.

Directs the Secretary to establish an innovative environmental technology transfer program to promote certain U.S. participation in international energy activities and the export of U.S. energy technologies. Prescribes implementation guidelines.

Requires the Secretary of the Treasury to establish a Global Climate Change Response Fund to serve as a conduit for U.S. contributions to global climate change. Authorizes appropriations.

Title XXIV: Non-Federal Power Act Hydropower Provisions—Prohibits FERC from issuing an original license for any new hydroelectric power project within the boundaries of any unit of the National Park System if it would have a direct adverse effect upon Federal lands within such unit.

Permits FERC to contract out statutorily mandated environmental impact statements. Prescribes guidelines under which the Secretary must study cost effective opportunities to increase hydropower production at existing federally-owned or operated water regulation, storage, and conveyance facilities.

Directs the Secretary of the Interior to conduct feasibility investigations of opportunities to: (1) increase certain hydroelectric energy available for marketing from Federal hydroelectric power generation facilities thanks to water conservation efforts on Federal reclamation projects; and (2) mitigate damages, or enhance fish and wildlife, through increasing available water. Authorizes appropriations.

Authorizes the Secretaries of the Interior and of the Army to implement generation additions, improvements and replacements at their respective Federal projects in the Pacific Northwest Region, and to operate their respective power facilities as needed, with the concurrence of the Bonneville Power Administrator.

Authorizes FERC to grant certain license requirement exemptions to itemized power projects in Alaska.

Requires FERC to study hydroelectric licensing in the State of Hawaii.

Title XXVIII: Nuclear Plant Licensing—Amends the Atomic Energy Act to prescribe conditions under which the NRC shall: (1) issue combined construction and operating licenses; and (2) hold post-construction hearings on such combined licenses. Subjects NRC determinations with respect to such licenses to judicial review.

APPENDIX C

U.S. STRATEGIES IN KEY SECTORS TO REDUCE NET EMISSIONS OF GREENHOUSE GASES

The following information was derived from the *U.S. Climate Action Report 2002*, which was written by the U.S. Department of State and published by the Government Printing Office.

Electricity: Federal programs promote greenhouse gas reductions through the development of cleaner, more efficient technologies for electricity generation and transmission. The government also supports the development of renewable resources, such as solar energy, wind power, geothermal energy, hydropower, bioenergy, and hydrogen fuels.

Transportation: Federal programs promote development of fuel-efficient motor vehicles and trucks, research and development options of producing cleaner fuels, and implementation of programs to reduce the number of vehicle miles traveled.

Industry: Federal programs implement partnership programs with industry to reduce emissions of carbon dioxide and other greenhouse gases, promote source reduction and recycling, and increase the use of combined heat and power.

Buildings: Federal voluntary partnership programs promote energy efficiency in the nation's commercial, residential, and government buildings (including schools) by offering technical assistance as well as the labeling of efficient products, new homes, and office buildings.

Agriculture and Forestry: The U.S. government implements conservation programs that have the benefit of reducing agricultural emissions, sequestering carbon in soils, and offsetting overall greenhouse gas emissions.

Federal Government: The U.S. Government has taken steps to reduce greenhouse gas emissions from energy use in federal buildings and in the federal transportation fleet.

APPENDIX D

U.S. CLIMATE ACTION REPORT, 2002, REDUCTION OF GREENHOUSE GASES

Summary of Actions to Reduce Greenhouse Gas Emissions

Name of Policy of Measure	Objective and/or Activity Affected	GHG Affected	Type of Instrument	Status	Implementing Entity/Entities
ENERGY: COMMERCIAL AND RESIDENTIAL					
Energy Star® for the commercial market	Promotes the improvement of energy performance in commercial buildings.	CO_2	Voluntary	Implemented	EPA
Commercial buildings integration: updating state buildings codes; partnerships for commercial buildings and facilities	Realizes energy-saving opportunities provided by whole-building approach during construction and major renovation of existing commercial buildings.	CO_2	Research, regulatory	Implemented	DOE
Energy Star® for the residential market	Promotes the improvement of energy performance in residential buildings beyond the labeling of products.	CO_2	Voluntary, outreach	Implemented	EPA
Community energy program: Rebuild America	Helps communities, towns and cities save energy, create jobs, promote growth, and protect the environment through improved energy efficiency and sustainable building design and operation.	CO_2	Voluntary, information, education	Implemented	DOE
Residential building integration: Building America	Funds, develops, demonstrates, and deploys housing that integrates energy-efficiency technologies and practices.	CO_2	Voluntary, research, education	Implemented	DOE
Energy Star®–Labeled Products	Label distinguishes energy-efficient products in the marketplace	CO_2	Voluntary, outreach	Implemented	EPA/DOE
Building Equipment, Materials, and Tools: Superwindow Collaborative; Lighting Partnerships; Partnerships for Commercial Buildings and Facilities; Collaborative Research and Development	Conducts research and development on building components and design tools and issues standards and test procedures for a variety of appliances and equipment.	CO_2	Information, research	Implemented	DOE

Residential Appliance Standards	Reviews and updates efficiency standards for most major household appliances.	CO_2	Regulatory	Implemented	DOE
State and Community Assistance: State Energy Program; Weatherization Assistance Program; Community Energy Grants; Information Outreach	Provides funding for state and communities to provide local energy-efficiency programs, including services to low-income families; to implement sustainable building design and operation; and to adopt a systematic approach to marketing and communication objectives.	CO_2	Economic, information	Implemented	DOE
Heat Island Reduction Initiative	Reverses the effects of urban heat islands by encouraging the use of mitigation strategies.	CO_2	Voluntary, information, research	Implemented	EPA
Economic Incentives/Tax Credits	Provides tax credits to residential solar energy systems.	CO_2	Economic	Proposed	
ENERGY: INDUSTRIAL					
Industries of the Future	Helps nine key energy-intensive industries reduce their energy consumption while remaining competitive and economically strong.	All	Voluntary, information	Implemented	DOE
Best Practices Program	Offers industry tools to improve plant energy efficiency, enhance environmental performance, and increase productivity.	All	Voluntary, information	Implemented	DOE
ENERGY STAR® for Industry (Climate Wise)	Enables industrial companies to evaluate and cost-effectively reduce their energy use.	CO_2	Voluntary	Implemented	EPA
Industrial Assessment Centers	Assesses and provides recommendations to manufacturers in identifying opportunities to improve productivity, reduce waste, and save energy.	All	Information, research	Implemented	DOE
Enabling Technologies	Addresses the critical technology challenges partners face for developing materials and production processes.	All	Information, research	Implemented	DOE

(Continued)

SUMMARY OF ACTIONS TO REDUCE GREENHOUSE GAS EMISSIONS (CONTINUED)

Name of Policy of Measure	Objective and/or Activity Affected	GHG Affected	Type of Instrument	Status	Implementing Entity/Entities
Financial Assistance: NICE[3]	Provides funding to state and industry partnerships for projects that develop and demonstrate advances in energy efficiency and clean production technologies.	All	Research	Implemented	DOE
ENERGY: SUPPLY					
Renewable Energy Commercialization: Wind; Solar; Geothermal; Biopower	Develops clean, competitive power technologies using renewable resources.	All	Research, regulatory	Implemented	DOE
Climate Challenge	Promotes efforts to reduce, avoid, or sequester greenhouse gases from electric utilities.	All	Voluntary	Implemented	DOE
Distributed Energy Resources (DER)	Focuses on technology development and the elimination of regulatory and institutional barriers to the use of DER.	All	Information, research, education, regulatory	Implemented	DOE
High-Temperature Superconductivity	Advances research and development of high-temperature superconducting power equipment for energy transmission, distribution, and industrial use.	All	Research	Implemented	DOE
Hydrogen Program	Enhances and supports the development of cost-competitive hydrogen technologies and systems to reduce the environmental impacts of their use.	All	Research, education	Implemented	DOE

Clean Energy Initiative: Green Power Partnership; Combined Heat and Power Partnership	Removes market barriers to increased penetration of cleaner, more efficient energy supply.	CO_2	Voluntary, education, technical assistance	Implemented	EPA
Nuclear Energy Plant Optimization	Recognizes the importance of existing nuclear plants in reducing greenhouse gas emissions.	CO_2	Information, technical assistance	Implemented	DOE
Development of Next-Generation Nuclear Energy Systems: Nuclear Energy Research Initiative; Generation IV Initiative	Supports research, development, and demonstration of an advanced nuclear energy system concept.	CO_2	Research, technical assistance	Implemented	DOE
Support Deployment of New Nuclear Power Plants in the United States	Ensures the availability of near-term nuclear energy options that can be in operation in the United States by 2010.	CO_2	Information	Implemented	DOE
Carbon Sequestration	Develops new technologies for addressing cost-effective management of CO_2 emissions from the production and use of fossil fuels.	CO_2	Research	Implemented	DOE
Hydropower Program	Improves the technical, societal, and environmental benefits of hydropower.	All	Information, research	Implemented	DOE
International Programs	Accelerates the international development and deployment of clean energy technologies.	All	Information, technical assistance	Implemented	DOE
Economic Incentives/Tax Credits	Provides tax credits to electricity generated from wind- and biomass-based generators.	CO_2	Economic	Proposed	
TRANSPORTATION					
FreedomCAR Research Partnership	Promotes the development of hydrogen as a primary fuel for cars and trucks.	CO_2	Research, information	Implemented	DOE
Vehicle Systems Research and Development	Promotes the development of cleaner, more efficient passenger vehicles.	CO_2	Research, information	Implemented	DOE

(Continued)

Summary of Actions to Reduce Greenhouse Gas Emissions (*Continued*)

Name of Policy of Measure	Objective and/or Activity Affected	GHG Affected	Type of Instrument	Status	Implementing Entity/Entities
Clean Cities	Supports public-private partnerships to deploy alternative-fuel vehicles and builds supporting infrastructure, including community networks.	All	Voluntary, information	Implemented	DOE
Biofuels Program	Researches, develops, demonstrates, and facilitates the commercialization of biomass-based, environmentally sound fuels for transportation.	All	Information, research	Implemented	DOE
Commuter Options Programs	Reduces single-occupant-vehicle commuting by providing incentives and alternative modes, timing, and locations for work.	CO_2	Voluntary agreements, tax incentives, information, education, outreach	Implemented	EPA/DOT
Smart Growth and Brownfields Policies	Reduces motorized trips and trip distance by promoting more efficient location choice.	CO_2	Technical assistance, outreach	Implemented	EPA
Ground Freight Transportation Initiative	Increases efficient management practices for ground freight.	CO_2	Voluntary/ negotiated agreements	Adopted	EPA
Clean Automotive Technology	Develops advanced clean and fuel-efficient automotive technology.	CO_2	Voluntary, research	Implemented	EPA
Department of Transportation Emission-Reducing Initiatives	Provides funding mechanisms for alternative modes to personal motorized vehicles.	CO_2	Funding mechanisms	Implemented	DOT
INDUSTRY (NON-CO2)					
Natural Gas STAR Program	Reduces methane emissions from U.S. natural gas systems through the wide-spread adoption of industry best management practices.	CH_4	Voluntary agreement	Implemented	EPA

Coalbed Methane Outreach Program	Reduces methane emissions from U.S. coal mining operations through cost-effective means.	CH_4	Information, education, outreach	Implemented	EPA
Significant New Alternatives Program	Facilitates smooth transition away from ozone-depleting chemicals in industrial and consumer sectors.	High GWP	Regulatory, information	Implemented	EPA
HFC-23 Partnership	Encourages reduction of HFC-23 emissions through cost-effective practices or technologies.	High GWP	Voluntary agreement	Implemented	EPA
Partnership with Aluminum Producers	Encourages reduction of CF_4 and C_2F_6 where technically feasible and cost-effective.	PFCs	Voluntary agreement	Implemented	EPA
Environmental Stewardship Initiative	Limits emissions of HFCs, PFCs, and SF_6 in industrial applications	High GWP	Voluntary agreement	Implemented	EPA
AGRICULTURE					
Agricultural Outreach Programs: AgSTAR; RLEP	Promotes practices to reduce GHG emissions at U.S. farms.	CH_4	Information education, outreach	Implemented	EPA/USDA
Nutrient Management Tools	Aims to reduce nitrous oxide emissions through improving by efficiency of fertilizer nitrogen.	N_2O	Technical assistance, information	Implemented	EPA/USDA
USDA CCC Bioenergy Program	Encourages bioenergy production through economic incentives to commodity producers.	CO_2	Economic	Implemented	USDA
Conservation Reserve Program: Biomass Project	Encourages land-use changes to increase the amount of feedstock available for biomass projects.	CO_2 N_2O	Economic	Implemented (pilot phase)	USDA
FORESTRY					
Forest Stewardship	Sequesters carbon in trees, forest soils, forest litter, and understudy plants.	CO_2	Technical/ financial assistance	Implemented	USDA

(Continued)

Summary of Actions to Reduce Greenhouse Gas Emissions (*Continued*)

Name of Policy of Measure	Objective and/or Activity Affected	GHG Affected	Type of Instrument	Status	Implementing Entity/Entities
WASTE MANAGEMENT					
Climate and Waste Program	Encourages recycling, source reduction, and other progressive integrated waste management activities to reduce GHG emissions.	All	Voluntary agreements, technical assistance, information, research	Implemented	EPA
Stringent Landfill Rule	Reduces methane/landfill gas emissions from U.S. landfills.	CH_4	Regulatory	Implemented	EPA
Landfill Methane Outreach Program	Reduces methane emissions from U.S. landfills through cost-effective means.	CH_4	Voluntary agreements, information, education, outreach	Implemented	EPA
CROSS-SECTORAL					
Federal Energy Management Program	Promotes energy efficiency and renewable energy use in federal buildings, facilities, and operations.	All	Economic, information, education	Implemented	DOE
State and Local Climate Change Outreach Program	Assists key state and local decision makers in maintaining and improving economic and environmental assets given climate change.	All	Information, education, research	Implemented	EPA

Source: U.S. Department of State. *U.S. Climate Action Report 2002*. Washington, D.C.: Government Printing Office, 2002.

APPENDIX E

EXCERPTS FROM GLOBAL WARMING AND EMISSIONS LEGISLATION IN THE U.S. CODE (2003)

Federal legislation is found in the U.S. Code. There are a number of provisions in the U.S. Code that relate to global warming and carbon dioxide emissions. Keep in mind that the U.S. Code is frequently updated by new legislation. The following excerpts are current as of mid-2003.

TITLE 15: COMMERCE AND TRADE

Sections 2901–2908 of Title 15 set the parameters of the National Climate Program, created by the National Climate Program Act of 1978 and amended in 2000. The program, designed to "assist the Nation and the world to understand and respond to natural and man-induced climate processes and their implications," includes provisions to increase climate research, improve forecasts, collect global data, disseminate data, and enhance international cooperation in all these areas. Sections 2921–2961 detail global change research policies enacted by the Global Change Research Act of 1990 to develop a research program designed to "assist the Nation and the world to understand, assess, predict, and respond to human-induced and natural processes of global change." The act called for the formation of the Committee on Earth Environmental Sciences, whose tasks include establishing goals and priorities for global change research, describing the steps necessary to achieve these goals, recommending ways that the United States can cooperate with other nations in achieving these goals, and preparing periodic scientific assessments to evaluate the findings of the program, analyze the effects of global change, and project major trends for the future.

15 U.S.C. SEC. 2901

Findings

The Congress finds and declares the following:

(1) Weather and climate change affect food production, energy use, land use, water resources and other factors vital to national security and human welfare.

(2) An ability to anticipate natural and man-induced changes in climate would contribute to the soundness of policy decisions in the public and private sectors.

(3) Significant improvements in the ability to forecast climate on an intermediate and long-term basis are possible.

(4) Information regarding climate is not being fully disseminated or used, and Federal efforts have given insufficient attention to assessing and applying this information.

(5) Climate fluctuation and change occur on a global basis, and deficiencies exist in the system for monitoring global climate changes. International cooperation for the purpose of sharing the benefits and costs of a global effort to understand climate is essential.

(6) The United States lacks a well-defined and coordinated program in climate-related research, monitoring, assessment of effects, and information utilization.

15 U.S.C. SEC. 2902

Purpose

It is the purpose of the Congress in this chapter to establish a national climate program that will assist the Nation and the world to understand and respond to natural and man-induced climate processes and their implications.

15 U.S.C. SEC. 2903

Definitions

As used in this chapter, unless the context otherwise requires:

(1) The term "Board" means the Climate Program Policy Board.

(2) The term "Office" means the National Climate Program Office.

(3) The term "Program" means the National Climate Program.
(4) The term "Secretary" means the Secretary of Commerce.

15 U.S.C. SEC. 2904

National Climate Program

(a) Establishment

The President shall establish a National Climate Program in accordance with the provisions, findings and purposes of this chapter.

(b) Duties

The President shall —

(1) promulgate the 5-year plans described in subsection (d) (9) of this section;

(2) define the roles in the Program of Federal officers, departments, and agencies, including the Departments of Agriculture, Commerce, Defense, Energy, Interior, State, and Transportation; the Environmental Protection Agency; the National Aeronautics and Space Administration; the Council on Environmental Quality; the National Science Foundation; and the Office of Science and Technology Policy; and

(3) provide for Program coordination.

(c) National Climate Program Office

(1) The Secretary shall establish within the Department of Commerce a National Climate Program Office not later than 30 days after September 17, 1978.

(2) The Office shall —

(A) serve as the lead entity responsible for administering the program;

(B) be headed by a Director who shall represent the Climate Program Policy Board and shall be spokesperson for the program;

(C) serve as the staff for the Board and its supporting committees and working groups;

(D) review each agency budget request transmitted under subsection (g) (1) of this section and submit an analysis of the requests to the Board for its review;

(E) be responsible for coordinating interagency participation in international climate-related activities; and

(F) work with the National Academy of Sciences and other private, academic, state, and local groups in preparing and implementing the 5-year plan (described in subsection (d) (9) of this section) and the program.

The analysis described in subparagraph (D) shall include an analysis of how each agency's budget request relates to the priorities and goals of the program established pursuant to this chapter.

(3) The Secretary may provide, through the Office, financial assistance, in the form of contracts or grants or cooperative agreements, for climate-related activities which are needed to meet the goals and priorities of the program set forth in the 5-year plan pursuant to subsection (d) (9) of this section, if such goals and priorities are not being adequately addressed by any Federal department, agency, or instrumentality.

(4) Each Federal officer, employee, department and agency involved in the Program shall cooperated with the Secretary in carrying out the provisions of this chapter.

(d) Program elements

The Program shall include, but not be limited to, the following elements:

(1) assessments of the effect of climate on the natural environment, agricultural production, energy supply and demand, land and water resources, transportation, human health and national security. Such assessments shall be conducted to the maximum extent possible by those Federal agencies having national programs in food, fiber, raw materials, energy, transportation, land and water management, and other such responsibilities, in accordance with existing laws and regulations. Where appropriate such assessments may include recommendations for action;

(2) basic and applied research to improve the understanding of climate processes, natural and man induced, and the social, economic, and political implications of climate change;

(3) methods for improving climate forecasts on a monthly, seasonal, yearly, and longer basis;

(4) global data collection, and monitoring and analysis activities to provide reliable, useful and readily available information on a continuing basis;

(5) systems for the management and active dissemination of climatological data, information and assessments, including mechanisms for consultation with current and potential users;

(6) measures for increasing international cooperation in climate research, monitoring, analysis and data dissemination;

(7) mechanisms for intergovernmental climate-related studies and services including participation by universities, the private sector and others concerned with applied research and advisory services. Such mechanisms may provide, among others, for the following State and regional services and functions:

(A) studies relating to and analyses of climatic effects on agricultural production, water resources, energy needs, and other critical sectors of the economy;

(B) atmospheric data collection and monitoring on a statewide and regional basis;

(C) advice to regional, State, and local government agencies regarding climate-related issues;

(D) information to users within the State regarding climate and climatic effects; and

(E) information to the Secretary regarding the needs of persons within the States for climate-related services, information, and data. The Secretary may make annual grants to any State or group of States, which grants shall be made available to public or private educational institutions, to State agencies, and to other persons or institutions qualified to conduct climate-related studies or provide climate-related services;

(8) experimental climate forecast centers, which shall

(A) be responsible for making and routinely updating experimental climate forecasts of a monthly, seasonal, annual, and longer nature, based on a variety of experimental techniques;

(B) establish procedures to have forecasts reviewed and their accuracy evaluated; and

(C) protect against premature reliance on such experimental forecasts; and

(9) a preliminary 5-year plan, to be submitted to the Congress for review and comment, not later than 180 days after September 17, 1978, and a final 5-year plan to be submitted to the Congress not later than 1 year after September 17, 1978, that shall be revised and extended at least once every four years. Each plan shall establish the goals and priorities for the Program, including the intergovernmental program described in paragraph (7), over the subsequent 5-year period, and shall contain details regarding

(A) the role of Federal agencies in the programs,

(B) Federal funding required to enable the Program to achieve such goals, and

(C) Program accomplishments that must be achieved to ensure that Program goals are met within the time frame established by the plan.

(e) Climate Program Policy Board

(1) The Secretary shall establish and maintain an interagency Climate Program Policy Board, consisting of representatives of the Federal agencies specified in subsection (b) (2) of this section and any other agency which the Secretary determines should participate in the Program.

(2) The Board shall —

(A) be responsible for coordinated planning and progress review for the Program;

(B) review all agency and department budget requests related to climate transmitted under subsection (g) (1) of this section and submit a report to the Office of Management and Budget concerning such budget requests;

(C) establish and maintain such interagency groups as the Board determines to be necessary to carry out its activities; and

(D) consult with and seek the advice of users and producers of climate data, information, and services to guide the Board's efforts, keeping the Director and the Congress advised of such contacts.

(3) The Board biennially shall select a Chair from among its members. A Board member who is a representative of an agency may not serve as Chair of the Board for a term if an individual who represented that same agency on the Board served as the Board's Chair for the previous term.

(f) Cooperation

(1) The Program shall be conducted so as to encourage cooperation with, and participation in the Program by, other organizations or agencies involved in related activities. For this purpose the Secretary shall cooperate and participate with other Federal agencies, and foreign, international, and domestic organizations and agencies involved in international or domestic climate-related programs.

(2) The Secretary and the Secretary of State shall cooperate with the Office in

(A) providing representation at climate-related international meetings and conferences in which the United States participates, and

(B) coordinating the activities of the Program with the climate programs of other nations and international agencies and organizations, including the World Meteorological Organization, the International Council of Scientific Unions, the United Nations Environmental Program, the United Nations Educational, Scientific, and Cultural Organization, the World Health Organization, and Food and Agriculture Organization.

(g) Budgeting Each Federal agency and department participating in the Program, shall prepare and submit to the Office of Management and Budget, on or before the date of submission of departmental requests for appropriations to the Office of Management and Budget, an annual request for appropriations for the Program for the subsequent fiscal year and shall transmit a copy of such request to the National Climate Program Office. The Office of Management and Budget shall review the request for appropriations as an integrated, coherent, multiagency request.

15 U.S.C. Sec. 2906

Annual report

The Secretary shall prepare and submit to the President and the authorizing committees of the Congress, not later than March 31 of each year, a report on the activities conducted pursuant to this chapter during the preceding fiscal year, including —

(a) a summary of the achievements of the Program during the previous fiscal year;

(b) an analysis of the progress made toward achieving the goals and objectives of the Program;

(c) a copy of the 5-year plan and any changes made in such plan;
(d) a summary of the multiagency budget request for the Program of section 2904(g) of this title; and
(e) any recommendations for additional legislation which may be required to assist in achieving the purposes of this chapter.

15 U.S.C. Sec. 2907

Contract and Grant Authority; Records and Audits

(a) Functions vested in any Federal officer or agency by this chapter or under the Program may be exercised through the facilities and personnel of the agency involved or, to the extent provided or approved in advance in appropriation Acts, by other persons or entities under contracts or grant arrangements entered into by such officer or agency.
(b)
(1) Each person or entity to which Federal funds are made available under a contract or grant arrangement as authorized by this chapter shall keep such records as the Director of the Office shall prescribe, including records which fully disclose the amount and disposition by such person or entity of such funds, the total cost of the activities for which such funds were so made available, the amount of that portion of such cost supplied from other sources, and such other records as will facilitate an effective audit.
(2) The Director of the Office and the Comptroller General of the United States, or any of their duly authorized representatives, shall, until the expiration of 3 years after the completion of the activities (referred to in paragraph (1)) of any person or entity pursuant to any contract or grant arrangement referred to in subsection (a) of this section, have access for the purpose of audit and examination to any books, documents, papers, and records of such person or entity which, in the judgment of the Director or the Comptroller General, may be related or pertinent to such contract or grant arrangement.

15 U.S.C. Sec. 2908

Authorization of appropriations

In addition to any other funds otherwise authorized to be appropriated for the purpose of conducting climate-related programs, there are authorized to be appropriated to the Secretary, for the purpose of carrying out the provisions of this chapter, not to exceed $50,000,000 for the fiscal year ending September 30, 1979, not to exceed $65,000,000 for the fiscal year ending

September 30, 1980, and not to exceed $25,500,000 for the fiscal year ending September 30, 1981, of which amount not less than $2,653,000 shall be made directly available to the National Climate Program Office in the form of a budget item separate from the activities of the National Oceanic and Atmospheric Administration.

Title 15, Sections 2921–2961 detail the policies of the United States concerning global change research as determined by the Global Change Research Act of 1990.

15 U.S.C. Sec. 2921

Definitions

As used in this chapter, the term —

(1) "Committee" means the Committee on Earth and Environmental Sciences established under section 2932 of this title;

(2) "Council" means the Federal Coordinating Council on Science, Engineering, and Technology;

(3) "global change" means changes in the global environment (including alterations in climate, land productivity, oceans or other water resources, atmospheric chemistry, and ecological systems) that may alter the capacity of the Earth to sustain life;

(4) "global change research" means study, monitoring, assessment, prediction, and information management activities to describe and understand —

(A) the interactive physical, chemical, and biological processes that regulate the total Earth system;

(B) the unique environment that the Earth provides for life;

(C) changes that are occurring in the Earth system; and

(D) the manner in which such system, environment, and changes are influenced by human actions;

(5) "Plan" means the National Global Change Research Plan developed under section 2934 of this title, or any revision thereof; and

(6) "Program" means the United States Global Change Research Program established under section 2933 of this title.

15 U.S.C. Sec. 2931

Findings and Purpose

(a) Findings

(b) The Congress makes the following findings:

(1) Industrial, agricultural, and other human activities, coupled with an expanding world population, are contributing to processes of global change that may significantly alter the Earth habitat within a few human generations.

(2) Such human-induced changes, in conjunction with natural fluctuations, may lead to significant global warming and thus alter world climate patterns and increase global sea levels. Over the next century, these consequences could adversely affect world agricultural and marine production, coastal habitability, biological diversity, human health, and global economic and social well-being.

(3) The release of chlorofluorocarbons and other stratospheric ozone-depleting substances is rapidly reducing the ability of the atmosphere to screen out harmful ultraviolet radiation, which could adversely affect human health and ecological systems.

(4) Development of effective policies to abate, mitigate, and cope with global change will rely on greatly improved scientific understanding of global environmental processes and on our ability to distinguish human-induced from natural global change.

(5) New developments in interdisciplinary Earth sciences, global observing systems, and computing technology make possible significant advances in the scientific understanding and prediction of these global changes and their effects.

(6) Although significant Federal global change research efforts are underway, an effective Federal research program will require efficient interagency coordination, and coordination with the research activities of State, private, and international entities.

(b) Purpose

The purpose of this subchapter is to provide for development and coordination of a comprehensive and integrated United States research program which will assist the Nation and the world to understand, assess, predict, and respond to human-induced and natural processes of global change.

15 U.S.C. Sec. 2932

Committee on Earth and Environmental Sciences

(a) Establishment

The President, through the Council, shall establish a Committee on Earth and Environmental Sciences. The Committee shall carry out Council functions under section 6651 of title 42 relating to global change research, for the purpose of increasing the overall effectiveness and productivity of Federal global change research efforts.

(b) Membership

The Committee shall consist of at least one representative from —

(1) the National Science Foundation;

(2) the National Aeronautics and Space Administration;

(3) the National Oceanic and Atmospheric Administration of the Department of Commerce;

(4) the Environmental Protection Agency;

(5) the Department of Energy;

(6) the Department of State;

(7) the Department of Defense;

(8) the Department of the Interior;

(9) the Department of Agriculture;

(10) the Department of Transportation;

(11) the Office of Management and Budget;

(12) the Office of Science and Technology Policy;

(13) the Council on Environmental Quality;

(14) the National Institute of Environmental Health Sciences of the National Institutes of Health; and

(15) such other agencies and departments of the United States as the President or the Chairman of the Council considers appropriate.

Such representatives shall be high ranking officials of their agency or department, wherever possible the head of the portion of that agency or department that is most relevant to the purpose of the subchapter described in section 2931(b) of this title.

(c) Chairperson

The Chairman of the Council, in consultation with the Committee, biennially shall select one of the Committee members to serve as Chairperson. The Chairperson shall be knowledgeable and experienced with regard to the administration of scientific research programs, and shall be a representative of an agency that contributes substantially, in terms of scientific research capability and budget, to the Program.

(d) Support personnel

An Executive Secretary shall be appointed by the Chairperson of the Committee, with the approval of the Committee. The Executive Secretary shall be a permanent employee of one of the agencies or departments represented on the Committee, and shall remain in the employ of such agency or department. The Chairman of the Council shall have the authority to make personnel decisions regarding any employees detailed to the Council for purposes of working on business of the Committee pursuant to section 6651 of title 42.

(e) Functions relative to global change

The Council, through the Committee, shall be responsible for planning and coordinating the Program. In carrying out this responsibility, the Committee shall —

(1) serve as the forum for developing the Plan and for overseeing its implementation;

(2) improve cooperation among Federal agencies and departments with respect to global change research activities;

(3) provide budgetary advice as specified in section 2935 of this title;

(4) work with academic, State, industry, and other groups conducting global change research, to provide for periodic public and peer review of the Program;

(5) cooperate with the Secretary of State in —

(A) providing representation at international meetings and conferences on global change research in which the United States participates; and

(B) coordinating the Federal activities of the United States with programs of other nations and with international global change research activities such as the International Geosphere-Biosphere Program;

(6) consult with actual and potential users of the results of the Program to ensure that such results are useful in developing national and international policy responses to global change; and

(7) report at least annually to the President and the Congress, through the Chairman of the Council, on Federal global change research priorities, policies, and programs.

15 U.S.C. Sec. 2933

United States Global Change Research Program

The President shall establish an interagency United States Global Change Research Program to improve understanding of global change. The Program shall be implemented by the Plan developed under section 2934 of this title.

15 U.S.C. Sec. 2934

National Global Change Research Plan

(a) In general

The Chairman of the Council, through the Committee, shall develop a National Global Change Research Plan for implementation of the Program. The Plan shall contain recommendations for national global change research. The Chairman of the Council shall submit the Plan to the Congress within one year after November 16, 1990, and a revised Plan shall be submitted at least once every three years thereafter.

(b) Contents of Plan

The Plan shall —

(1) establish, for the 10-year period beginning in the year the Plan is submitted, the goals and priorities for Federal global change research which most effectively advance scientific understanding of global change and provide usable information on which to base policy decisions relating to global change;

(2) describe specific activities, including research activities, data collection and data analysis requirements, predictive modeling, participation in international research efforts, and information management, required to achieve such goals and priorities;

(3) identify and address, as appropriate, relevant programs and activities of the Federal agencies and departments represented on the Committee that contribute to the Program;

(4) set forth the role of each Federal agency and department in implementing the Plan;

(5) consider and utilize, as appropriate, reports and studies conducted by Federal agencies and departments, the National Research Council, or other entities;

(6) make recommendations for the coordination of the global change research activities of the United States with such activities of other nations and international organizations, including—

(A) a description of the extent and nature of necessary international cooperation;

(B) the development by the Committee, in consultation when appropriate with the National Space Council, of proposals for cooperation on major capital projects;

(C) bilateral and multilateral proposals for improving worldwide access to scientific data and information; and

(D) methods for improving participation in international global change research by developing nations; and

(7) estimate, to the extent practicable, Federal funding for global change research activities to be conducted under the Plan.

(c) Research elements

The Plan shall provide for, but not be limited to, the following research elements:

(1) Global measurements, establishing worldwide observations necessary to understand the physical, chemical, and biological processes responsible for changes in the Earth system on all relevant spatial and time scales.

(2) Documentation of global change, including the development of mechanisms for recording changes that will actually occur in the Earth system over the coming decades.

(3) Studies of earlier changes in the Earth system, using evidence from the geological and fossil record.

(4) Predictions, using quantitative models of the Earth system to identify and simulate global environmental processes and trends, and the regional implications of such processes and trends.

(5) Focused research initiatives to understand the nature of and interaction among physical, chemical, biological, and social processes related to global change.

(d) Information management

The Plan shall provide recommendations for collaboration within the Federal Government and among nations to —

(1) establish, develop, and maintain information bases, including necessary management systems which will promote consistent, efficient, and compatible transfer and use of data;

(2) create globally accessible formats for data collected by various international sources; and

(3) combine and interpret data from various sources to produce information readily usable by policymakers attempting to formulate effective strategies for preventing, mitigating, and adapting to the effects of global change.

(e) National Research Council evaluation

The Chairman of the Council shall enter into an agreement with the National Research Council under which the National Research Council shall —

(1) evaluate the scientific content of the Plan; and

(2) provide information and advice obtained from United States and international sources, and recommended priorities for future global change research.

(f) Public participation

In developing the Plan, the Committee shall consult with academic, State, industry, and environmental groups and representatives. Not later than 90 days before the Chairman of the Council submits the Plan, or any revision thereof, to the Congress, a summary of the proposed Plan shall be published in the Federal Register for a public comment period of not less than 60 days.

15 U.S.C. Sec. 2935

Budget Coordination

(a) Committee guidance

The Committee shall each year provide general guidance to each Federal agency or department participating in the Program with respect to the preparation of requests for appropriations for activities related to the Program.

(b) Submission of reports with agency appropriations requests

(1) Working in conjunction with the Committee, each Federal agency or department involved in global change research shall include with its annual

request for appropriations submitted to the President under section 1108 of title 31 a report which —

(A) identifies each element of the proposed global change research activities of the agency or department;

(B) specifies whether each element

(i) contributes directly to the Program or

(ii) contributes indirectly but in important ways to the Program; and

(C) states the portion of its request for appropriations allocated to each element of the Program.

(2) Each agency or department that submits a report under paragraph (1) shall submit such report simultaneously to the Committee.

(c) Consideration in President's budget

(1) The President shall, in a timely fashion, provide the Committee with an opportunity to review and comment on the budget estimate of each agency and department involved in global change research in the context of the Plan.

(2) The President shall identify in each annual budget submitted to the Congress under section 1105 of title 31 those items in each agency's or department's annual budget which are elements of the Program.

15 U.S.C. Sec. 2936

Scientific Assessment

On a periodic basis (not less frequently than every 4 years), the Council, through the Committee, shall prepare and submit to the President and the Congress an assessment which —

(1) integrates, evaluates, and interprets the findings of the Program and discusses the scientific uncertainties associated with such findings;

(2) analyzes the effects of global change on the natural environment, agriculture, energy production and use, land and water resources, transportation, human health and welfare, human social systems, and biological diversity; and

(3) analyzes current trends in global change, both human-induced and natural, and projects major trends for the subsequent 25 to 100 years.

15 U.S.C. Sec. 2938

Relation to Other Authorities

(a) National Climate Program research activities

The President, the Chairman of the Council, and the Secretary of Commerce shall ensure that relevant research activities of the National Climate

Program, established by the National Climate Program Act (15 U.S.C. 2901 et seq.), are considered in developing national global change research efforts.

(b) Availability of research findings

The President, the Chairman of the Council, and the heads of the agencies and departments represented on the Committee, shall ensure that the research findings of the Committee, and of Federal agencies and departments, are available to —

(1) the Environmental Protection Agency for use in the formulation of a coordinated national policy on global climate change pursuant to section 1103 of the Global Climate Protection Act of 1987 (15 U.S.C. 2901 note); and

(2) all Federal agencies and departments for use in the formulation of coordinated national policies for responding to human-induced and natural processes of global change pursuant to other statutory responsibilities and obligations.

(c) Effect on Federal response actions

Nothing in this subchapter shall be construed, interpreted, or applied to preclude or delay the planning or implementation of any Federal action designed, in whole or in part, to address the threats of stratospheric ozone depletion or global climate change.

15 U.S.C. SEC. 2951

Findings and Purposes

(a) Findings

(b) The Congress makes the following findings:

(1) Pooling of international resources and scientific capabilities will be essential to a successful international global change program.

(2) While international scientific planning is already underway, there is currently no comprehensive intergovernmental mechanism for planning, coordinating, or implementing research to understand global change and to mitigate possible adverse effects.

(3) An international global change research program will be important in building future consensus on methods for reducing global environmental degradation.

(4) The United States, as a world leader in environmental and Earth sciences, should help provide leadership in developing and implementing an international global change research program.

(b) Purposes

The purposes of this subchapter are to —

(1) promote international, intergovernmental cooperation on global change research;

(2) involve scientists and policymakers from developing nations in such cooperative global change research programs; and

(3) promote international efforts to provide technical and other assistance to developing nations which will facilitate improvements in their domestic standard of living while minimizing damage to the global or regional environment.

15 U.S.C. SEC. 2952

International Discussions

(a) Global change research

The President should direct the Secretary of State, in cooperation with the Committee, to initiate discussions with other nations leading toward international protocols and other agreements to coordinate global change research activities. Such discussions should include the following issues:

(1) Allocation of costs in global change research programs, especially with respect to major capital projects.

(2) Coordination of global change research plans with those developed by international organizations such as the International Council on Scientific Unions, the World Meteorological Organization, and the United Nations Environment Program.

(3) Establishment of global change research centers and training programs for scientists, especially those from developing nations.

(4) Development of innovative methods for management of international global change research, including —

(A) use of new of existing intergovernmental organizations for the coordination or funding of global change research; and

(B) creation of a limited foundation for global change research.

(5) The prompt establishment of international projects to —

(A) create globally accessible formats for data collected by various international sources; and

(B) combine and interpret data from various sources to produce information readily usable by policymakers attempting to formulate effective strategies for preventing, mitigating, and adapting to possible adverse effects of global change.

(6) Establishment of international offices to disseminate information useful in identifying, preventing, mitigating, or adapting to the possible effects of global change.

(b) Energy research

The President should direct the Secretary of State (in cooperation with the Secretary of Energy, the Secretary of Commerce, the United States

Trade Representative, and other appropriate members of the Committee) to initiate discussions with other nations leading toward an international research protocol for cooperation on the development of energy technologies which have minimally adverse effects on the environment. Such discussions should include, but not be limited to, the following issues:

(1) Creation of an international cooperative program to fund research related to energy efficiency, solar and other renewable energy sources, and passively safe and diversion-resistant nuclear reactors.

(2) Creation of an international cooperative program to develop low cost energy technologies which are appropriate to the environmental, economic, and social needs of developing nations.

(3) Exchange of information concerning environmentally safe energy technologies and practices, including those described in paragraphs (1) and (2).

15 U.S.C. Sec. 2953

Global Change Research Information Office

Not more than 180 days after November 16, 1990, the President shall, in consultation with the Committee and all relevant Federal agencies, establish an Office of Global Change Research Information. The purpose of the Office shall be to disseminate to foreign governments, businesses, and institutions, as well as the citizens of foreign countries, scientific research information available in the United States which would be useful in preventing, mitigating, or adapting to the effects of global change. Such information shall include, but need not be limited to, results of scientific research and development on technologies useful for —

(1) reducing energy consumption through conservation and energy efficiency;

(2) promoting the use of solar and renewable energy sources which reduce the amount of greenhouse gases released into the atmosphere;

(3) developing replacements for chlorofluorocarbons, halons, and other ozone-depleting substances which exhibit a significantly reduced potential for depleting stratospheric ozone;

(4) promoting the conservation of forest resources which help reduce the amount of carbon dioxide in the atmosphere;

(5) assisting developing countries in ecological pest management practices and in the proper use of agricultural, and industrial chemicals; and

(6) promoting recycling and source reduction of pollutants in order to reduce the volume of waste which must be disposed of, thus decreasing energy use and greenhouse gas emissions.

15 U.S.C. SEC. 2961

Study and Decision Aid

(a) Study of consequences of community growth and development; decision aid to assist State and local authorities in managing development

The Secretary of Commerce shall conduct a study of the implications and potential consequences of growth and development on urban, suburban, and rural communities. Based upon the findings of the study, the Secretary shall produce a decision aid to assist State and local authorities in planning and managing urban, suburban, and rural growth and development while preserving community character.

(b) Consultation with appropriate Federal departments and agencies

The Secretary of Commerce shall consult with other appropriate Federal departments and agencies as necessary in carrying out this section.

(c) Report

The Secretary of Commerce shall submit to the Congress a report containing the decision aid produced under subsection (a) of this section no later than January 30, 1992. The Secretary shall notify appropriate State and local authorities that such decision aid is available on request

The purpose of this subchapter is to provide for development and coordination of a comprehensive and integrated United States research program which will assist the Nation and the world to understand, assess, predict, and respond to human-induced and natural processes of global change.

TITLE 16: CONSERVATION

Section 4501 authorizes international forestry activities for the purpose of reducing the impact of global greenhouse gas emissions. Provisions focus on promoting sustainable development and global environmental stability, sharing natural resource skills, providing education and training, cooperating in research, and cooperating with organizations that focus on protection of natural resources.

16 U.S.C. SEC. 4501

Forestry and Related Natural Resource Assistance

(a) Focus of activities

To achieve the maximum impact from activities undertaken under the authority of this chapter, the Secretary shall focus such activities on the key countries which could have a substantial impact on emissions of greenhouse gases related to global warming.

(b) Authority for international forestry activities

In support of forestry and related natural resource activities outside of the United States and its territories and possessions, the Secretary of Agriculture may —

(1) provide assistance that promotes sustainable development and global environmental stability, including assistance for —

(A) conservation and sustainable management of forest land;

(B) forest plantation technology and tree improvement;

(C) rehabilitation of cutover lands, eroded watersheds, and areas damaged by wildfires or other natural disasters;

(D) prevention and control of insects, diseases, and other damaging agents;

(E) preparedness planning, training, and operational assistance to combat natural disasters;

(F) more complete utilization of forest products leading to resource conservation;

(G) range protection and enhancement; and

(H) wildlife and fisheries habitat protection and improvement;

(2) share technical, managerial, extension, and administrative skills related to public and private natural resource administration;

(3) provide education and training opportunities to promote the transfer and utilization of scientific information and technologies;

(4) engage in scientific exchange and cooperative research with foreign governmental, educational, technical and research institutions; and

(5) cooperate with domestic and international organizations that further international programs for the management and protection of forests, rangelands, wildlife and fisheries, and related natural resource activities.

(c) Eligible countries

The Secretary shall undertake the activities described in subsection (b) of this section, in countries that receive assistance from the Agency for International Development only at the request, or with the concurrence, of the Administrator of the Agency for International Development.

TITLE 22: FOREIGN RELATIONS AND INTERCOURSE

Section 2431 details provisions to protect tropical forests around the world, acknowledged as "playing a critical role as carbon sinks in reducing greenhouse gases in the atmosphere," by alleviating debt in nations where such forests are located. This relief will allow governments in these nations to allocate additional money to protect their natural resources.

22 U.S.C. Sec. 2431

Debt Reduction for Developing Countries with Tropical Forests Findings and Purpose

(a) Findings

The Congress finds the following:

(1) It is the established policy of the United States to support and seek protection of tropical forests around the world.

(2) Tropical forests provide a wide range of benefits to humankind by —

(A) harboring a major share of the Earth's biological and terrestrial resources, which are the basis for developing pharmaceutical products and revitalizing agricultural crops;

(B) playing a critical role as carbon sinks in reducing greenhouse gases in the atmosphere, thus moderating potential global climate change; and

(C) regulating hydrological cycles on which far-flung agricultural and coastal resources depend.

(3) International negotiations and assistance programs to conserve forest resources have proliferated over the past decade, but the rapid rate of tropical deforestation continues unabated.

(4) Developing countries with urgent needs for investment and capital for development have allocated a significant amount of their forests to logging concessions.

(5) Poverty and economic pressures on the populations of developing countries have, over time, resulted in clearing of vast areas of forest for conversion to agriculture, which is often unsustainable in the poor soils underlying tropical forests.

(6) Debt reduction can reduce economic pressures on developing countries and result in increased protection for tropical forests.

(7) Finding economic benefits to local communities from sustainable uses of tropical forests is critical to the protection of tropical forests.

(b) Purposes

The purposes of this subchapter are —

(1) to recognize the values received by United States citizens from protection of tropical forests;

(2) to facilitate greater protection of tropical forests (and to give priority to protecting tropical forests with the highest levels of biodiversity and under the most severe threat) by providing for the alleviation of debt in countries where tropical forests are located, thus allowing the use of additional resources to protect these critical resources and reduce economic pressures that have led to deforestation;

(3) to ensure that resources freed from debt in such countries are targeted to protection of tropical forests and their associated values; and

(4) to rechannel existing resources to facilitate the protection of tropical forests.

TITLE 42: THE PUBLIC HEALTH AND WELFARE

The energy policy of the United States is detailed in Sections 13201–13556. Provisions relevant to global warming studies include the implementation of the Replacement Fuel Supply and Demand Program, designed to promote the replacement of petroleum-based motor fuels with alternative fuels; a least-cost energy strategy that promotes more efficient use of existing energy technologies and the development of renewable energy sources; a national inventory of greenhouse gas emissions to help determine the best ways to deal with the global warming problem; a technology transfer program to promote the export of U.S. energy technologies and technological expertise; and the Global Climate Change Response Fund as a means to assist global efforts to mitigate and adapt to climate change.

42 U.S.C. SEC. 13252

Replacement Fuel Supply and Demand Program

(a) Establishment of program

The Secretary shall establish a program to promote the development and use in light duty motor vehicles of domestic replacement fuels. Such program shall promote the replacement of petroleum motor fuels with replacement fuels to the maximum extent practicable. Such program shall, to the extent practicable, ensure the availability of those replacement fuels that will have the greatest impact in reducing oil imports, improving the health of our Nation's economy and reducing greenhouse gas emissions.

(b) Development plan and production goals Under the program established under subsection (a) of this section, the Secretary, before October 1, 1993, in consultation with the Administrator, the Secretary of Transportation, the Secretary of Agriculture, the Secretary of Commerce, and the heads of other appropriate agencies, shall review appropriate information and —

(1) estimate the domestic and nondomestic production capacity for replacement fuels and alternative fueled vehicles needed to implement this section;

(2) determine the technical and economic feasibility of achieving the goals of producing sufficient replacement fuels to replace, on an energy equivalent basis —

(A) at least 10 percent by the year 2000; and

(B) at least 30 percent by the year 2010, of the projected consumption of motor fuel in the United States for each such year, with at least one half of such replacement fuels being domestic fuels;

(3) determine the most suitable means and methods of developing and encouraging the production, distribution, and use of replacement fuels and alternative fueled vehicles in a manner that would meet the program goals described in subsection (a) of this section;

(4) identify ways to encourage the development of reliable replacement fuels and alternative fueled vehicle industries in the United States, and the technical, economic, and institutional barriers to such development; and

(5) determine the greenhouse gas emission implications of increasing the use of replacement fuels, including an estimate of the maximum feasible reduction in such emissions from the use of replacement fuels. The Secretary shall publish in the Federal Register the results of actions taken under this subsection, and provide for an opportunity for public comment.

42 U.S.C. Sec. 13253

Replacement Fuel Demand Estimates and Supply Information

(a) Estimates

Not later than October 1, 1993, and annually thereafter, the Secretary, in consultation with the Administrator, the Secretary of Transportation, and other appropriate State and Federal officials, shall estimate for the following calendar year —

(1) the number of each type of alternative fueled vehicle likely to be in use in the United States;

(2) the probable geographic distribution of such vehicles;

(3) the amount and distribution of each type of replacement fuel; and

(4) the greenhouse gas emissions likely to result from replacement fuel use.

(b) Information

Beginning on October 1, 1994, the Secretary shall annually require —

(1) fuel suppliers to report to the Secretary on the amount of each type of replacement fuel that such supplier —

(A) has supplied in the previous calendar year; and

(B) plans to supply for the following calendar year;

(2) suppliers of alternative fueled vehicles to report to the Secretary on the number of each type of alternative fueled vehicle that such supplier —

(A) has made available in the previous calendar year; and

(B) plans to make available for the following calendar year; and

(3) such fuel suppliers to provide the Secretary information necessary to determine the greenhouse gas emissions from the replacement fuels used, taking into account the entire fuel cycle.

(c) Protection of information

Information provided to the Secretary under subsection (b) of this section shall be subject to applicable provisions of law protecting the confidentiality of trade secrets and business and financial information, including section 1905 of title 18.

42 U.S.C. Sec. 13315

Data System and Energy Technology Evaluation

The Secretary of Commerce, in his or her role as a member of the interagency working group established under section 6276 of this title, shall —

(1) develop a comprehensive data base and information dissemination system, using the National Trade Data Bank and the Commercial Information Management System of the Department of Commerce, that will provide information on the specific energy technology needs of foreign countries, and the technical and economic competitiveness of various renewable energy and energy efficiency products and technologies;

(2) make such information available to industry, Federal and multilateral lending agencies, nongovernmental organizations, host-country and donor-agency officials, and such others as the Secretary of Commerce considers necessary; and

(3) prepare and transmit to the Congress not later than June 1, 1993, and biennially thereafter, a comprehensive report evaluating the full range of energy and environmental technologies necessary to meet the energy needs of foreign countries, including —

(A) information on the specific energy needs of foreign countries;

(B) an inventory of United States technologies and services to meet those needs;

(C) an update on the status of ongoing bilateral and multilateral programs which promote United States exports of renewable energy and energy efficiency products and technologies; and

(D) an evaluation of current programs (and recommendations for future programs) that develop and promote energy efficiency and sustainable use of indigenous renewable energy resources in foreign countries to reduce the generation of greenhouse gases.

42 U.S.C. SEC. 13331

Coal Research, Development, Demonstration, and Commercial Application Programs

(a) Establishment

The Secretary shall, in accordance with section [1] 13541 and 13542 of this title, conduct programs for research, development, demonstration, and commercial application on coal-based technologies. Such research, development, demonstration, and commercial application programs shall include the programs established under this part, and shall have the goals and objectives of—

(1) ensuring a reliable electricity supply;

(2) complying with applicable environmental requirements;

(3) achieving the control of sulfur oxides, oxides of nitrogen, air toxics, solid and liquid wastes, greenhouse gases, or other emissions resulting from coal use or conversion at levels of proficiency greater than or equal to applicable currently available commercial technology;

(4) achieving the cost competitive conversion of coal into energy forms usable in the transportation sector;

(5) demonstrating the conversion of coal to synthetic gaseous, liquid, and solid fuels;

(6) demonstrating, in cooperation with other Federal and State agencies, the use of coal-derived fuels in mobile equipment, with opportunities for industrial cost sharing participation;

(7) ensuring the timely commercial application of cost-effective technologies or energy production processes or systems utilizing coal which achieve—

(A) greater efficiency in the conversion of coal to useful energy when compared to currently available commercial technology for the use of coal; and

(B) the control of emissions from the utilization of coal; and

(8) ensuring the availability for commercial use of such technologies by the year 2010.

(b) Demonstration and commercial application programs

(1) In selecting either a demonstration project or a commercial application project for financial assistance under this part, the Secretary shall seek to ensure that, relative to otherwise comparable commercially available technologies or products, the selected project will meet one or more of the following criteria:

(A) It will reduce environmental emissions to an extent greater than required by applicable provisions of law.

(B) It will increase the overall efficiency of the utilization of coal, including energy conversion efficiency and, where applicable, production of products derived from coal.

(C) It will be a more cost-effective technological alternative, based on life cycle capital and operating costs per unit of energy produced and, where applicable, costs per unit of product produced.

Priority in selection shall be given to those projects which, in the judgment of the Secretary, best meet one or more of these criteria.

(2) In administering demonstration and commercial application programs authorized by this part, the Secretary shall establish accounting and project management controls that will be adequate to control costs.

(3)

(A) Not later than 180 days after October 24, 1992, the Secretary shall establish procedures and criteria for the recoupment of the Federal share of each cost shared demonstration and commercial application project authorized pursuant to this part. Such recoupment shall occur within a reasonable period of time following the date of completion of such project, but not later than 20 years following such date, taking into account the effect of recoupment on —

(i) the commercial competitiveness of the entity carrying out the project;

(ii) the profitability of the project; and

(iii) the commercial viability of the coal-based technology utilized.

(B) The Secretary may at any time waive or defer all or some portion of the recoupment requirement as necessary for the commercial viability of the project.

(4) Projects selected by the Secretary under this part for demonstration or commercial application of a technology shall, in the judgment of the Secretary, be capable of enhancing the state of the art for such technology.

(c) Report

Within 240 days after October 24, 1992, the Secretary shall transmit to the Committee on Energy and Commerce and the Committee on Science, Space, and Technology of the House of Representatives and to the Committee on Energy and Natural Resources of the Senate a report which shall include each of the following:

(1) A detailed description of ongoing research, development, demonstration, and commercial application activities regarding coal-based technologies undertaken by the Department of Energy, other Federal or State government departments or agencies and, to the extent such information is publicly available, other public or private organizations in the United States and other countries.

(2) A listing and analysis of current Federal and State government regulatory and financial incentives that could further the goals of the programs established under this part.

(3) Recommendations regarding the manner in which any ongoing coal-based demonstration and commercial application program might be modi-

fied and extended in order to ensure the timely demonstrations of advanced coal-based technologies so as to ensure that the goals established under this section are achieved and that such demonstrated technologies are available for commercial use by the year 2010.

(4) Recommendations, if any, regarding the manner in which the cost sharing demonstrations conducted pursuant to the Clean Coal Program established by Public Law 98–473 might be modified and extended in order to ensure the timely demonstration of advanced coal-based technologies.

(5) A detailed plan for conducting the research, development, demonstration, and commercial application programs to achieve the goals and objectives of subsection

(a) of this section, which plan shall include a description of—

(A) the program elements and management structure to be utilized;

(B) the technical milestones to be achieved with respect to each of the advanced coal-based technologies included in the plan; and

(C) the dates at which further deadlines for additional cost sharing demonstrations shall be established.

(d) Status reports

Within one year after transmittal of the report described in subsection (c) of this section, and every 2 years thereafter for a period of 6 years, the Secretary shall transmit to the Congress a report that provides a detailed description of the status of development of the advanced coal-based technologies and the research, development, demonstration, and commercial application activities undertaken to carry out the programs required by this part.

(e) Consultation

In carrying out research, development, demonstration, and commercial application activities under this part, the Secretary shall consult with the National Coal Council and other representatives of the public and private sectors as the Secretary considers appropriate.

42 U.S.C. Sec. 13381

Report

Not later than 2 years after October 24, 1992, the Secretary shall submit a report to the Congress that includes an assessment of—

(1) the feasibility and economic, energy, social, environmental, and competitive implications, including implications for jobs, of stabilizing the generation of greenhouse gases in the United States by the year 2005;

(2) the recommendations made in chapter 9 of the 1991 National Academy of Sciences report entitled "Policy Implications of Greenhouse Warming," including an analysis of the benefits and costs of each recommendation;

(3) the extent to which the United States is responding, compared with other countries, to the recommendations made in chapter 9 of the 1991 National Academy of Sciences report;

(4) the feasibility of reducing the generation of greenhouse gases;

(5) the feasibility and economic, energy, social, environmental, and competitive implications, including implications for jobs, of achieving a 20 percent reduction from 1988 levels in the generation of carbon dioxide by the year 2005 as recommended by the 1988 Toronto Scientific World Conference on the Changing Atmosphere;

(6) the potential economic, energy, social, environmental, and competitive implications, including implications for jobs, of implementing the policies necessary to enable the United States to comply with any obligations under the United Nations Framework Convention on Climate Change or subsequent international agreements.

42 U.S.C. Sec. 13382

Least-Cost Energy Strategy

(a) Strategy

The first National Energy Policy Plan (in this subchapter referred to as the "Plan") under section 7321 of this title prepared and required to be submitted by the President to Congress after February 1, 1993, and each subsequent such Plan, shall include a least-cost energy strategy prepared by the Secretary. In developing the least-cost energy strategy, the Secretary shall take into consideration the economic, energy, social, environmental, and competitive costs and benefits, including costs and benefits for jobs, of his choices. Such strategy shall also take into account the report required under section 13381 of this title and relevant Federal, State, and local requirements. Such strategy shall be designed to achieve to the maximum extent practicable and at least-cost to the Nation —

(1) the energy production, utilization, and energy conservation priorities of subsection (d) of this section;

(2) the stabilization and eventual reduction in the generation of greenhouse gases;

(3) an increase in the efficiency of the Nation's total energy use by 30 percent over 1988 levels by the year 2010;

(4) an increase in the percentage of energy derived from renewable resources by 75 percent over 1988 levels by the year 2005; and

(5) a reduction in the Nation's oil consumption from the 1990 level of approximately 40 percent of total energy use to 35 percent by the year 2005.

(b) Additional contents

The least-cost energy strategy shall also include —

(1) a comprehensive inventory of available energy and energy efficiency resources and their projected costs, taking into account all costs of production, transportation, distribution, and utilization of such resources, including —

(A) coal, clean coal technologies, coal seam methane, and underground coal gasification;

(B) energy efficiency, including existing technologies for increased efficiency in production, transportation, distribution, and utilization of energy, and other technologies that are anticipated to be available through further research and development; and

(C) other energy resources, such as renewable energy, solar energy, nuclear fission, fusion, geothermal, biomass, fuel cells, hydropower, and natural gas;

(2) a proposed two-year program for ensuring adequate supplies of the energy and energy efficiency resources and technologies described in paragraph (1), and an identification of administrative actions that can be undertaken within existing Federal authority to ensure their adequate supply;

(3) estimates of life-cycle costs for existing energy production facilities;

(4) basecase forecasts of short-term and long-term national energy needs under low and high case assumptions of economic growth; and

(5) an identification of all applicable Federal authorities needed to achieve the purposes of this section, and of any inadequacies in those authorities.

(c) Secretarial consideration

In developing the least-cost energy strategy, the Secretary shall give full consideration to —

(1) the relative costs of each energy and energy efficiency resource based upon a comparison of all direct and quantifiable net costs for the resource over its available life, including the cost of production, transportation, distribution, utilization, waste management, environmental compliance, and, in the case of imported energy resources, maintaining access to foreign sources of supply; and

(2) the economic, energy, social, environmental, and competitive consequences resulting from the establishment of any particular order of Federal priority as determined under subsection (d) of this section.

(d) Priorities

The least-cost energy strategy shall identify Federal priorities, including policies that —

(1) implement standards for more efficient use of fossil fuels;

(2) increase the energy efficiency of existing technologies;

(3) encourage technologies, including clean coal technologies, that generate lower levels of greenhouse gases;

(4) promote the use of renewable energy resources, including solar, geothermal, sustainable biomass, hydropower, and wind power;

(5) affect the development and consumption of energy and energy efficiency resources and electricity through tax policy;

(6) encourage investment in energy efficient equipment and technologies; and

(7) encourage the development of energy technologies, such as advanced nuclear fission and nuclear fusion, that produce energy without greenhouse gases as a byproduct, and encourage the deployment of nuclear electric generating capacity.

(e) Assumptions

The Secretary shall include in the least-cost energy strategy an identification of all of the assumptions used in developing the strategy and priorities thereunder, and the reasons for such assumptions.

(f) Preference

When comparing an energy efficiency resource to an energy resource, a higher priority shall be assigned to the energy efficiency resource whenever all direct and quantifiable net costs for the resource over its available life are equal to the estimated cost of the energy resource.

(g) Public review and comment

The Secretary shall provide for a period of public review and comment of the least-cost energy strategy, for a period of at least 30 days, to be completed at least 60 days before the issuance of such strategy. The Secretary shall also provide for public review and comment before the issuance of any update to the least-cost energy strategy required under this section.

42 U.S.C. Sec. 13383

Director of Climate Protection

Within 6 months after October 24, 1992, the Secretary shall establish, within the Department of Energy, a Director of Climate Protection (in this section referred to as the "Director"). The Director shall —

(1) in the absence of the Secretary, serve as the Secretary's representative for interagency and multilateral policy discussions of global climate change, including the activities of the Committee on Earth and Environmental Sciences as established by the Global Change Research Act of 1990 (Public Law 101–606) (15 U.S.C. 2921 et seq.) and the Policy Coordinating Committee Working Group on Climate Change;

(2) monitor, in cooperation with other Federal agencies, domestic and international policies for their effects on the generation of greenhouse gases; and

(3) have the authority to participate in the planning activities of relevant Department of Energy programs.

42 U.S.C. SEC. 13384

Assessment of Alternative Policy Mechanisms for Addressing Greenhouse Gas Emissions

Not later than 18 months after October 24, 1992, the Secretary shall transmit a report to Congress containing a comparative assessment of alternative policy mechanisms for reducing the generation of greenhouse gases. Such assessment shall include a short-run and long-run analysis of the social, economic, energy, environmental, competitive, and agricultural costs and benefits, including costs and benefits for jobs and competition, and the practicality of each of the following policy mechanisms:

(1) Various systems for controlling the generation of greenhouse gases, including caps for the generation of greenhouse gases from major sources and emissions trading programs.

(2) Federal standards for energy efficiency for major sources of greenhouse gases, including efficiency standards for power plants, industrial processes, automobile fuel economy, appliances, and buildings, and for emissions of methane.

(3) Various Federal and voluntary incentives programs.

42 U.S.C. SEC. 13385

National Inventory and Voluntary Reporting of Greenhouse Gases

(a) National inventory

Not later than one year after October 24, 1992, the Secretary, through the Energy Information Administration, shall develop, based on data available to, and obtained by, the Energy Information Administration, an inventory of the national aggregate emissions of each greenhouse gas for each calendar year of the baseline period of 1987 through 1990. The Administrator of the Energy Information Administration shall annually update and analyze such inventory using available data. This subsection does not provide any new data collection authority.

(b) Voluntary reporting

(1) Issuance of guidelines

Not later than 18 months after October 24, 1992, the Secretary shall, after opportunity for public comment, issue guidelines for the voluntary collection and reporting of information on sources of greenhouse gases. Such guidelines shall establish procedures for the accurate voluntary reporting of information on —

(A) greenhouse gas emissions —

(i) for the baseline period of 1987 through 1990; and

(ii) for subsequent calendar years on an annual basis;

(B) annual reductions of greenhouse gas emissions and carbon fixation achieved through any measures, including fuel switching, forest management practices, tree planting, use of renewable energy, manufacture or use of vehicles with reduced greenhouse gas emissions, appliance efficiency, energy efficiency, methane recovery, cogeneration, chlorofluorocarbon capture and replacement, and power plant heat rate improvement;

(C) reductions in greenhouse gas emissions achieved as a result of—

(i) voluntary reductions;

(ii) plant or facility closings; and

(iii) State or Federal requirements; and

(D) an aggregate calculation of greenhouse gas emissions by each reporting entity.

Such guidelines shall also establish procedures for taking into account the differential radiative activity and atmospheric lifetimes of each greenhouse gas.

(2) Reporting procedures

The Administrator of the Energy Information Administration shall develop forms for voluntary reporting under the guidelines established under paragraph (1), and shall make such forms available to entities wishing to report such information. Persons reporting under this subsection shall certify the accuracy of the information reported.

(3) Confidentiality

Trade secret and commercial or financial information that is privileged or confidential shall be protected as provided in section 552(b)(4) of title 5.

(4) Establishment of data base

Not later than 18 months after October 24, 1992, the Secretary, through the Administrator of the Energy Information Administration, shall establish a data base comprised of information voluntarily reported under this subsection. Such information may be used by the reporting entity to demonstrate achieved reductions of greenhouse gases.

(c) Consultation

In carrying out this section, the Secretary shall consult, as appropriate, with the Administrator of the Environmental Protection Agency.

42 U.S.C. Sec. 13387

Innovative Environmental Technology Transfer Program

(a) Establishment of program

The Secretary, through the Agency for International Development, and in consultation with the interagency working group established under section 6276(d) of this title (in this section referred to as the "interagency working group," [1] shall establish a technology transfer program to carry out the purposes described in subsection (b) of this section. Within 150 days after October 24, 1992, the Secretary and the Administrator of the Agency for International Development shall enter into a written agreement to carry out this section. The agreement shall establish a procedure for resolving any disputes between the Secretary and the Administrator regarding the implementation of specific projects. With respect to countries not assisted by the Agency for International Development, the Secretary may enter into agreements with other appropriate Federal agencies. If the Secretary and the Administrator, or the Secretary and an agency described in the previous sentence, are unable to reach an agreement, each shall send a memorandum to the President outlining an appropriate agreement. Within 90 days after receipt of either memorandum, the President shall determine which version of the agreement shall be in effect. Any agreement entered into under this subsection shall be provided to the appropriate committees of the Congress and made available to the public.

(b) Purposes of program

The purposes of the technology transfer program under this section are to —

(1) reduce the United States balance of trade deficit through the export of United States energy technologies and technological expertise;

(2) retain and create manufacturing and related service jobs in the United States;

(3) encourage the export of United States technologies, including services related thereto, to those countries that have a need for developmentally sound facilities to provide energy derived from technologies that substantially reduce environmental pollutants, including greenhouse gases;

(4) develops markets for United States technologies, including services related thereto, that substantially reduce environmental pollutants, including greenhouse gases, that meet the energy and environmental requirements of foreign countries;

(5) better ensure that United States participation in energy-related projects in foreign countries includes participation by United States firms as well as utilization of United States technologies;

(6) ensure the introduction of United States firms and expertise in foreign countries;

(7) provide financial assistance by the Federal Government to foster greater participation by United States firms in the financing, ownership, design, construction, or operation of technologies or services that substantially reduce environmental pollutants, including greenhouse gases; and

(8) assist United States firms, especially firms that are in competition with firms in foreign countries, to obtain opportunities to transfer technologies to, or undertake projects in, foreign countries.

(c) Identification

Pursuant to the agreements required by subsection (a) of this section, the Secretary, through the Agency for International Development, and after consultation with the interagency working group, United States firms, and representatives from foreign countries, shall develop mechanisms to identify potential energy projects in host countries that substantially reduce environmental pollutants, including greenhouse gases, and shall identify a list of such projects within 240 days after October 24, 1992, and periodically thereafter.

(d) Financial mechanisms

(1) Pursuant to the agreements under subsection (a) of this section, the Secretary, through the Agency for International Development, shall —

(A) establish appropriate financial mechanisms to increase the participation of United States firms in energy projects, and services related thereto, that substantially reduce environmental pollutants, including greenhouse gases in foreign countries;

(B) utilize available financial assistance authorized by this section to counterbalance assistance provided by foreign governments to non-United States firms; and

(C) provide financial assistance to support projects.

(2) The financial assistance authorized by this section may be —

(A) provided in combination with other forms of financial assistance, including non-Federal funding that may be available for the project; and

(B) utilized in conjunction with financial assistance programs available through other Federal agencies.

(3) United States obligations under the Arrangement on Guidelines for Officially Supported Export Credits established through the Organization for Economic Cooperation and Development shall be applicable to this section.

(e) Solicitations for project proposals

(1) Pursuant to the agreements under subsection (a) of this section, the Secretary, through the Agency for International Development, within one year after October 24, 1992, and subsequently as appropriate thereafter, shall solicit proposals from United States firms for the design, construction, testing, and operation of the project or projects identified under subsection (c) of this section which propose to utilize a United States technology or service. Each solicitation under this section shall establish a closing date for receipt of proposals.

(2) The solicitation under this subsection shall, to the extent appropriate, be modeled after the RFP No. DE-PS01-90FE62271 Clean Coal Technology IV, as administered by the Department of Energy.

(3) Any solicitation made under this subsection shall include the following requirements:

(A) The United States firm that submits a proposal in response to the solicitation shall have an equity interest in the proposed project.

(B) The project shall utilize a United States technology, including services related thereto, that substantially reduce environmental pollutants, including greenhouse gases, in meeting the applicable energy and environmental requirements of the host country.

(C) Proposals for projects shall be submitted by and undertaken with a United States firm, although a joint venture or other teaming arrangement with a non-United States manufacturer or other non-United States entity is permissible.

(f) Assistance to United States firms

Pursuant to the agreements under subsection (a) of this section, the Secretary, through the Agency for International Development, and in consultation with the interagency working group, shall establish a procedure to provide financial assistance to United States firms under this section for a project identified under subsection (c) of this section where solicitations for the project are being conducted by the host country or by a multilateral lending institution.

(g) Other program requirements

Pursuant to the agreements under subsection (a) of this section, the Secretary, through the Agency for International Development, and in consultation with the interagency working group, shall —

(1) establish eligibility criteria for countries that will host projects;

(2) periodically review the energy needs of such countries and export opportunities for United States firms for the development of projects in such countries;

(3) consult with government officials in host countries and, as appropriate, with representatives of utilities or other entities in host countries, to determine interest in and support for potential projects; and

(4) determine whether each project selected under this section is developmentally sound, as determined under the criteria developed by the Development Assistance Committee of the Organization for Economic Cooperation and Development.

(h) Eligible technologies

Not later than 6 months after October 24, 1992, the Secretary shall prepare a list of eligible technologies and services under this section. In preparing such a list, the Secretary shall consider fuel cell power plants, aeroderivative gas turbines and catalytic combustion technologies for aeroderivitive gas turbines, ocean thermal energy conversion technology, anaerobic digester and storage tanks, and other renewable energy and energy efficiency technologies.

(i) Selection of projects

(1) Pursuant to the agreements under subsection (a) of this section, the Secretary, through the Agency for International Development, shall, not later than 120 days after receipt of proposals in response to a solicitation under subsection (e) of this section, select one or more proposals under this section.

(2) In selecting a proposal under this section, the Secretary, through the Agency for International Development, shall consider —

(A) the ability of the United States firm, in cooperation with the host country, to undertake and complete the project;

(B) the degree to which the equipment to be included in the project is designed and manufactured in the United States;

(C) the long-term technical and competitive viability of the United States technology, and services related thereto, and the ability of the United States firm to compete in the development of additional energy projects using such technology in the host country and in other foreign countries;

(D) the extent of technical and financial involvement of the host country in the project;

(E) the extent to which the proposed project meets the purposes of this section;

(F) the extent of technical, financial, management, and marketing capabilities of the participants in the project, and the commitment of the participants to completion of a successful project in a manner that will facilitate acceptance of the United States technology or service for future application; and

(G) such other criteria as may be appropriate.

(3) In selecting among proposed projects, the Secretary shall seek to ensure that, relative to otherwise comparable projects in the host country, a selected project will meet the following criteria:

(A) It will reduce environmental emissions, including greenhouse gases, to an extent greater than required by applicable provisions of law.

(B) It will be a more cost-effective technological alternative, based on life cycle capital and operating costs per unit of energy produced and, where applicable, costs per unit of product produced.

(C) It will increase the overall efficiency of energy use. Priority in selection shall be given to those projects which, in the judgment of the Secretary, best meet these criteria.

(j) United States-Asia Environmental Partnership Activities carried out under this section shall be coordinated with the United States-Asia Environmental Partnership.

(k) Buy America

In carrying out this section, the Secretary, through the Agency for International Development, and pursuant to the agreements under subsection (a) of this section, shall ensure —

(1) the maximum percentage, but in no case less than 50 percent, of the cost of any equipment furnished in connection with a project authorized under this section shall be attributable to the manufactured United States components of such equipment; and

(2) the maximum participation of United States firms. In determining whether the cost of United States components equals or exceeds 50 percent, the cost of assembly of such United States components in the host country shall not be considered a part of the cost of such United States component.

(1) Report to Congress

The Secretary and the Administrator of the Agency for International Development shall report annually to the Committee on Energy and Natural Resources of the Senate and the appropriate committees of the House of Representatives on the progress being made to introduce innovative energy technologies, and services related thereto, that substantially reduce environmental pollutants, including greenhouse gases, into foreign countries.

(m) Definitions

For purposes of this section —

(1) the term "host country" means a foreign country which is —

(A) the participant in or the site of the proposed innovative energy technology project; and

(B) either —

(i) classified as a country eligible to participate in development assistance programs of the Agency for International Development pursuant to applicable law or regulation; or

(ii) a developing country; and

(2) the term "developing country" includes, but is not limited to, countries in Central and Eastern Europe or in the independent states of the former Soviet Union.

(n) Authorization of appropriations

There are authorized to be appropriated to the Secretary to carry out the program required by this section, $100,000,000 for each of the fiscal years 1993, 1994, 1995, 1996, 1997, and 1998.

42 U.S.C. Sec. 13388

Global Climate Change Response Fund

(a) Establishment of Fund

The Secretary of the Treasury, in consultation with the Secretary of State, shall establish a Global Climate Change Response Fund to act as a mechanism for United States contributions to assist global efforts in mitigating and adapting to global climate change.

(b) Restrictions on deposits

No deposits shall be made to the Global Climate Change Response Fund until the United States has ratified the United Nations Framework Convention on Climate Change.

(c) Use of Fund

Moneys deposited into the Fund shall be used by the President, to the extent authorized and appropriated under section 2222 of title 22, solely for contributions to a financial mechanism negotiated pursuant to the United Nations Framework Convention on Climate Change, including all protocols or agreements related thereto.

(d) Authorization of appropriations

There are authorized to be appropriated for deposit in the Fund to carry out the purposes of this section, $50,000,000 for fiscal year 1994 and such sums as may be necessary for fiscal years 1995 and 1996.

42 U.S.C. Sec. 13401

Goals

It is the goal of the United States in carrying out energy supply and energy conservation research and development —

(1) to strengthen national energy security by reducing dependence on imported oil;

(2) to increase the efficiency of the economy by meeting future needs for energy services at the lowest total cost to the Nation, including environmental costs, giving comparable consideration to technologies that enhance energy supply and technologies that improve the efficiency of energy end uses;

(3) to reduce the air, water, and other environmental impacts (including emissions of greenhouse gases) of energy production, distribution, transportation, and utilization, through the development of an environmentally sustainable energy system;

(4) to maintain the technological competitiveness of the United States and stimulate economic growth through the development of advanced materials and technologies;

(5) to foster international cooperation by developing international markets for domestically produced sustainable energy technologies, and by transferring environmentally sound, advanced energy systems and technologies to developing countries to promote sustainable development;

(6) to consider the comparative environmental and public health impacts of the energy to be produced or saved by the specific activities;

(7) to consider the obstacles inherent in private industry's development of new energy technologies and steps necessary for establishing or maintaining technological leadership in the area of energy and energy efficiency resource technologies; and

(8) to consider the contribution of a given activity to fundamental scientific knowledge.

APPENDIX F

CLIMATE CHANGE

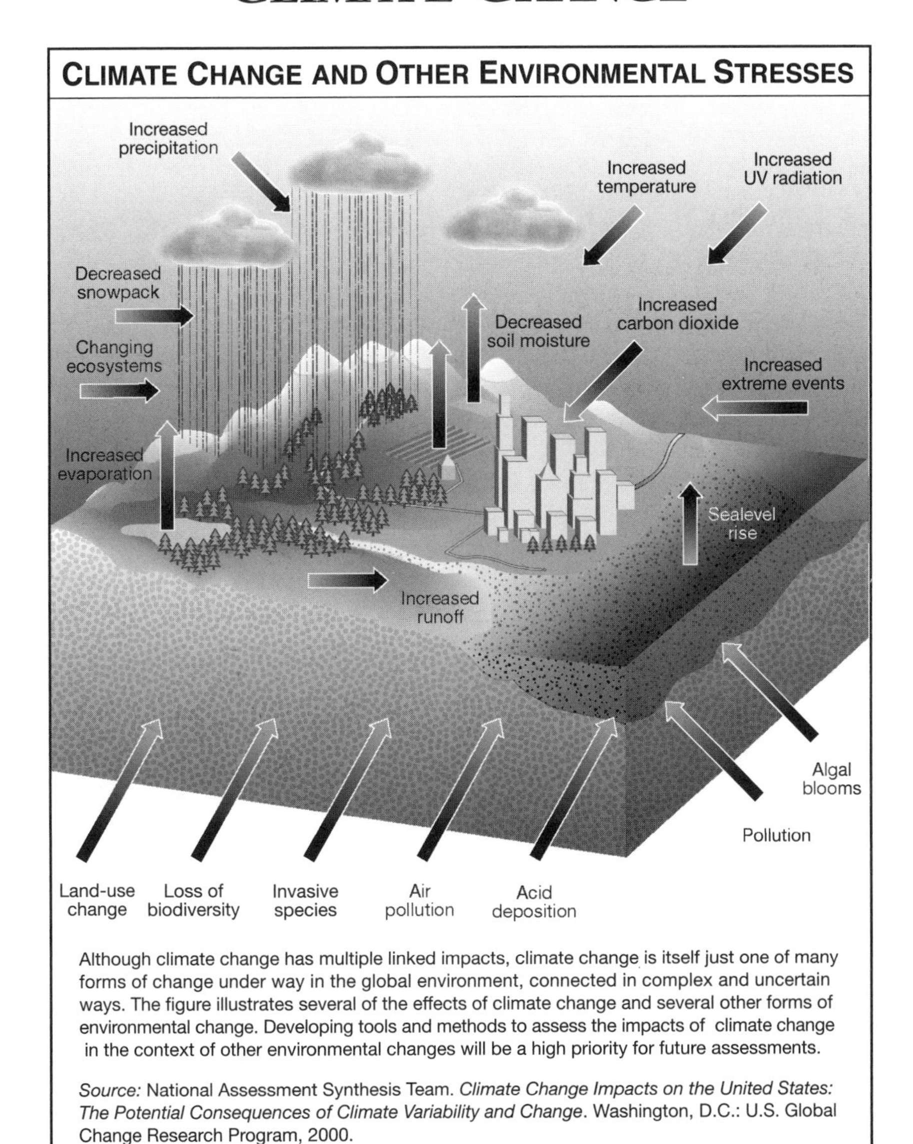

CLIMATE CHANGE AND OTHER ENVIRONMENTAL STRESSES

Although climate change has multiple linked impacts, climate change is itself just one of many forms of change under way in the global environment, connected in complex and uncertain ways. The figure illustrates several of the effects of climate change and several other forms of environmental change. Developing tools and methods to assess the impacts of climate change in the context of other environmental changes will be a high priority for future assessments.

Source: National Assessment Synthesis Team. *Climate Change Impacts on the United States: The Potential Consequences of Climate Variability and Change*. Washington, D.C.: U.S. Global Change Research Program, 2000.

APPENDIX G

EXCERPT FROM U.S. CLIMATE ACTION REPORT, 2002

CONSEQUENCES OF GLOBAL WARMING, KEY REGIONAL VULNERABILITY AND CONSEQUENCE ISSUES

The following key vulnerability and consequence issues were identified across the setoff regions considered in the U.S. National Assessment. Additional details may be found in the regional reports indexed at http://www.usgcrp.gov.

Northeast, Southeast, and Midwest: Rising temperatures are likely to increase the heat index dramatically in summer. Warmer winters are likely to reduce cold-related stresses. Both types of changes are likely to affect health and comfort.

Appalachians: Warmer and moister air is likely to lead to more intense rainfall events in mountainous areas, increasing the potential for flash floods.

Great Lakes: Lake levels are likely to decline due to increased warm-season evaporation, leading to reduced water supply and degraded water quality. Lower lake levels are also likely to increase shipping costs, although a longer shipping season is likely. Shoreline damage due to high water levels is likely to decrease, but reduced wintertime ice cover is likely to lead to higher waves and greater shoreline erosion.

Southeast: Under warmer, wetter scenarios, the range of southern tree species is likely to expand. Under hotter, drier scenarios, it is likely that grasslands and savannas will eventually displace southeastern forests in many areas, with transformation likely accelerated by increased occurrence of large fires.

Appendix G

Southeast Atlantic Coast, Puerto Rico, and the Virgin Islands: Rising sea level and higher storm surges are likely to cause loss of many coastal ecosystems that now provide an important buffer for coastal development against the impacts of storms. Currently and newly exposed communities are more likely to suffer damage from increasing intensity of storms.

Midwest/Great Plains: A rising carbon dioxide concentration is likely to offset the effects of rising temperatures on forests and agriculture for several decades, increasing productivity and thereby reducing commodity prices for the public. To the extent that overall production is not increased, higher crop and forest productivity is likely to lead to less land being farmed and logged, which may promote recovery of some natural environments.

Great Plains: Prairie potholes, which provide important habitats for ducks and other migratory waterfowl, are likely to become much drier in a warmer climate.

Southwest: With an increase in precipitation, the desert ecosystems native to this region are likely to be replaced in many areas by grasslands and shrublands, increasing both fire and agricultural potential.

Northern and Mountain Regions: It is very likely that warm-weather recreational opportunities like hiking will expand, while cold-weather activities like skiing will contract.

Mountain West: Higher winter temperatures are very likely to reduce late winter snow pack. This is likely to cause peak runoff to be lower, which is likely to reduce the potential for spring floods associated with snowmelt. As the peak flow shifts to earlier in the spring, summer runoff is likely to be reduced, which is likely to require modifications in water management to provide for flood control, power production, fish runs, cities, and irrigation.

Northwest: Increasing river and stream temperatures are very likely to further stress migrating fish, complicating current restoration efforts.

Alaska: Sharp winter and springtime temperature increases are very likely to cause continued melting of sea ice and thawing of permafrost, further disrupting ecosystems, infrastructure, and communities. A longer warm season could also increase opportunities for shipping, commerce, and tourism.

Hawaii and Pacific Trust Territories: More intense El Niño and La Niña events are possible and would be likely to create extreme fluctuations in water resources for island citizens and the tourists who sustain local economies.

Source: U.S. Department of State. *U.S. Climate Action Report 2002.* Washington, D.C.: Government Printing Office, 2002.

APPENDIX H

GLOBAL WARMING RESEARCH

FUNDAMENTAL CLIMATE CHANGE RESEARCH NEEDS

Research Uncertainty	U.S. Global Change Research Program Research Focus
ATMOSPHERIC COMPOSITION	
• How do human activities and natural phenomena change the composition of the global atmosphere? • How do these changes influence climate, ozone, ultraviolet radiation, pollutant exposure, ecosystems, and human health?	• Processes affecting the recovery of the stratospheric ozone layer. • Properties and distribution of greenhouse gases and aerosols. • Long-range transport of pollutants and implications for air quality. • Integrated assessments of the effects of these changes for the nation and the world.
CLIMATE VARIABILITY AND CHANGE	
• How do changes in the Earth system that result from natural processes and human activities affect the climate elements that are important to human and natural systems, especially temperature, precipitation, clouds, winds, and extreme events?	• Predictions of seasonal-to-decadal climate variations (*e.g.,* the El Niño-Southern Oscillation). • Detection and attribution of human-induced change. • Projections of long-term climate change. • Potential for changes in extreme events at regional-to-local scales. • Possibility of abrupt climate change. • How to improve the effectiveness of interactions between producers and users of climate forecast information.
CARBON CYCLE	
• How large and variable are the reservoirs and transfers of carbon within the Earth system?	• North American and ocean carbon sources and sinks. • Impacts of land-use changes and resource management practices on carbon sources and sinks.

FUNDAMENTAL CLIMATE CHANGE RESEARCH NEEDS

Research Uncertainty	U.S. Global Change Research Program Research Focus
• How might carbon sources and sinks change and be managed in the future?	• Future atmospheric carbon dioxide and methane concentrations and changes in land-based and marine carbon sinks. • Periodic reporting (starting in 2010) on the global distribution of carbon sources and sinks and how they are changing.
GLOBAL WATER CYCLE	
• How do human activities and natural processes that affect climate variability influence the distribution and quality of water within the Earth system? • To what extent are these changes predictable? • How will these changes affect climate, the cycling of carbon and other nutrients, and other environmental properties?	• Trends in the intensity of the water cycle and the causes of these changes (including feedback effects of clouds on the water and energy budgets, as well as the global climate system). • Predictions of precipitation and evaporation on time scales of months to years and longer. • Models of physical and biological processes and human demands and institutional processes, to facilitate efficient management of water resources. • Research supporting reports on the state of the global water cycle and national water resources.
TERRESTRIAL AND MARINE ECOSYSTEMS	
• How do natural and human-induced changes in the environment interact to affect ecosystems (from natural to intensively managed), their ability to provide natural resources and commodities, and their influence on regional and global climate?	• Structure and function of ecosystems, including cycling of nutrients and how they interact with the carbon cycle. • Key processes that link ecosystems with climate. • Vulnerability of ecosystems to global change. • Options for enhancing resilience and sustaining ecosystem goods and services. • Scientific underpinning for improved interactions with resource managers.
CHANGES IN LAND USE AND LAND COVER	
• What processes determine land cover and land use at local, regional, and global scales? • How will land use and land cover evolve over time scales of 10–50 years?	• Identifying the human drivers of changes in land use and cover. • Monitoring, measuring, and mapping land use and land cover and managing data systems. • Developing projections of land-cover and land-use changes under various assumptions about climate, demographic, economic, and technological trends. • Integrating information about land use, land management, and land cover into other research elements.

Note: To support informed decision making, the U.S. Global Change Research Program is addressing uncertainties about how human activities are changing the Earth's climate and environment. The six research elements in this table focus on topics essential to projecting climate change and understanding its potential importance.

Source: U.S. Department of State. *U.S. Climate Action Report 2002.* Washington, D.C.: Government Printing Officer, 2002.

MODELING THE CLIMATE

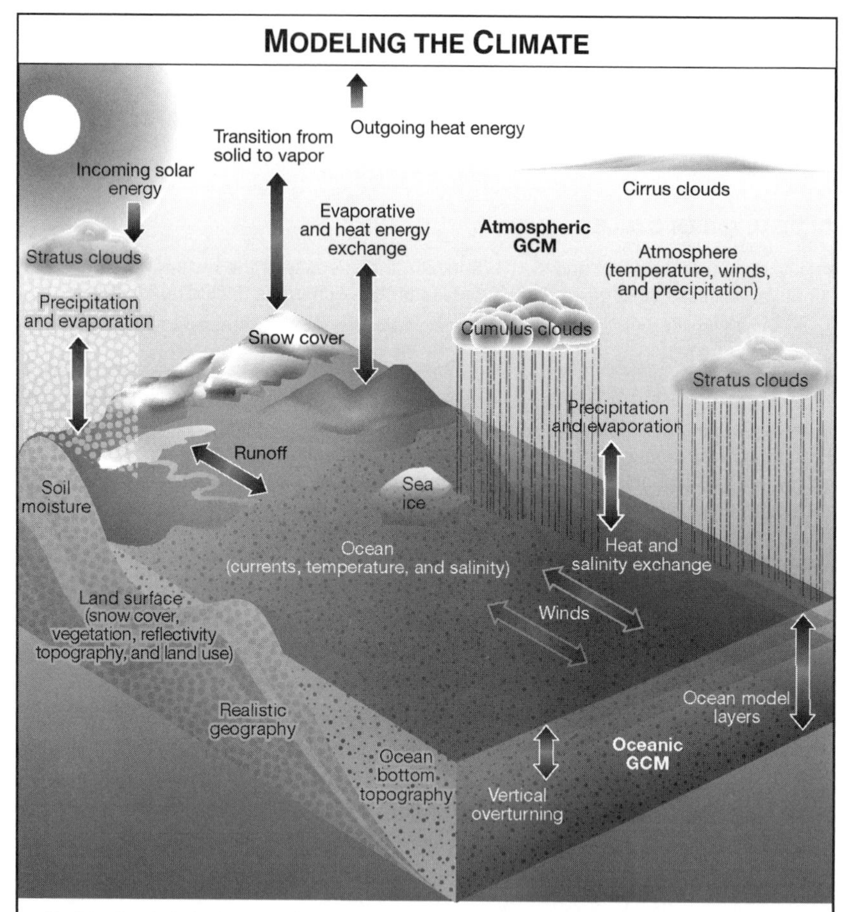

Earth's climate is far too complex to reproduce in a laboratory. An alternative is to devise a mathematical representation, or model, that can be used to simulate past, present, and future climate conditions. These models incorporate the key physical parameters and processes that govern climate behavior. Once constructed, they can be used to investigate how a change in greenhouse gases, or a volcanic eruption, might modify the climate.

Computer models that simulate Earth's climate are called general circulation models (GCMs). The models can be used to simulate changes in temperature, rainfall, snow cover, winds, soil moisture, sea ice, and ocean circulation over the entire globe through the seasons and over periods of decades. However, mathematical models are obviously simplified versions of the real Earth that cannot capture its full complexity, especially at smaller geographic scales. Real uncertainties remain in the ability of models to simulate many aspects of the future climate. The models provide a view of future climate that is physically consistent and plausible but incomplete. Nonetheless, through continual improvement over the last several decades, today's GCMs provide a state-of-the-science glimpse into the next century to help understand how climate may affect the nation.

Source: National Assessment Synthesis Team. *Climate Change Impacts on the United States: The Potential Consequences of Climate Variability and Change*. Washington, D.C.: U.S. Global Change Research Program, 2000.

Index

Page numbers in **boldface** indicate main topics. Page numbers followed by *g* indicate glossary entries. Page numbers followed by *c* indicate chronology entries. Page numbers followed by *b* indicate biographical entries.

Index

Index

Index

Index

Index

Index